Contents

Appendixes 77

Preface

This revision of *Laboratory methods for sensory evaluation of food* has been updated throughout to include changes and advances that have occurred since the first edition was published in 1967 and revised in 1977.

Because of continued demand for the publication, especially as a teaching aid and for use in small companies with limited technical expertise, an updated version was felt to be necessary.

Three dedicated young scientists who now lead the sensory analysis team at Agriculture Canada, Ottawa, have made significant contributions in the updating and revision. It has been a pleasure to work with them and to pass on the gauntlet to a new generation.

This edition is still intended as a manual and is prescriptive in nature. Readers are referred to some of the excellent texts that deal with the principles underlying the practices recommended here.

Elizabeth Larmond

Acknowledgments

The authors thank John Wiley and Sons, Inc., for permission to reprint in part the *Tables of random permutations* (Statistical Chart 1) from *Experimental designs* by William G. Cochran and Gertrude M. Cox, 1957. We would also like to thank the Institute of Food Technologists for permission to reprint statistical charts 2, 3, and 4 from the *Journal of Food Science* 43(3):942–943 by E.B. Roessler, R.M. Pangborn, J.L. Sidel, and H. Stone (copyright by the Institute of Food Technologists). We acknowledge with thanks CRC Press Inc. for permission to reprint Table T5 (Statistical Chart 6) from *Sensory analysis techniques, Volume II* by M. Meilgaard, G. Vance Civille, and Thomas B. Carr, 1987. We thank Iowa State University Press, Ames, Iowa, for permission to reprint statistical charts 7 and 8 from *Statistical methods* by George W. Snedecor and William G. Cochran, 8th edition, 1989. We are also grateful to the literary executor of the late Sir Ronald A. Fisher, FRS, Cambridge, Dr. Frank Yates, FRS, Rothamstead, and Messrs. Oliver and Boyd Ltd., Edinburgh, for permission to reprint in part Table V (Statistical Chart 9) from their book *Statistical tables for biological, agricultural, and medical research* (6th edition 1974). The authors also thank the Biometric Society for permission to reproduce statistical charts 10 and 11. We also wish to thank Diversified Research Laboratories Limited, Toronto, for permission to use a photograph of their facilities (Fig. 1). We owe a special debt of gratitude to everyone who has advised and worked with us in preparing this book and Drs. J.J. Powers, M. McDaniel, and B.K. Thompson for their reviews. Finally, our most sincere thanks and gratitude go to Mrs. Josiane Obas for her dedicated efforts in typing our manuscript.

Introduction

Sensory evaluation was defined by the Sensory Evaluation Division of the Institute of Food Technologists (1975) as "the scientific discipline used to evoke, measure, analyze and interpret those reactions to characteristics of foods and materials as perceived through the senses of sight, smell, taste, touch and hearing." The complex sensation that results from the interaction of our senses is used to measure food quality in programs such as quality control and new product development. This evaluation may be carried out by panels of a small number of people or by several hundred depending on the type of information required.

The first and simplest form of sensory analysis is made at the bench by the research worker who develops the new food products or quality control specifications. Researchers rely on their own evaluation to determine gross differences in products. Sensory analysis is conducted in a more formal manner by laboratory and consumer panels.

Most sensory characteristics of food can only be measured well, completely, and meaningfully by human subjects. Advances continue to be made in developing instrumental tests that measure individual quality factors. As instruments are developed to measure these factors, sensory analysis data are correlated with the results to determine their predictive ability.

When people are used as a measuring instrument, it is necessary to control all testing methods and conditions rigidly to overcome errors caused by psychological factors. "Error" is not synonymous with mistakes but may include all kinds of extraneous influences. The physical and mental condition of the panelists and the influence of the testing environment affect their sensory responses.

Sensory analysis panels can be grouped into four types: highly trained experts, trained laboratory panels, laboratory acceptance panels, and large consumer panels.

Highly trained experts (1–3 people) evaluate quality with a very high degree of acuity and reproducibility, e.g., wine, tea, and coffee experts. Evaluations by experts and trained laboratory panels can be useful for control purposes, for guiding product development and improvement, and for evaluating quality. The trained panel (10–20 people) can be particularly useful in assessing product attribute changes for which there is no adequate instrumentation. Sensory analyses performed by laboratory acceptance panels (25–50 people) are valuable in predicting consumer reactions to a product. Large consumer panels (more than 100 people) are used to determine consumer reaction to a product.

Factors influencing sensory measurements

Standard procedures for planning and conducting sensory panels have been developed in an effort to minimize or control the effect that psychological errors and physical conditions can have on human judgment. The need for standardized procedures can perhaps be emphasized by describing some of the factors that affect human judgment and by outlining ways in which to minimize or eliminate them.

Expectation error

Any information that panelists receive about the test can influence the results. Panelists usually find what they expect to find. Therefore, give panelists only enough information for them to conduct the test. Do not include on the panel those persons who are directly involved with the experiment. Code the samples so that the panelists cannot identify them, as the code itself should introduce no bias. Because people generally associate "1" or "A" with "best," we recommend the use of three-digit random numbers.

Stimulus error

In a desire to be right, the judgment of the panel members may be influenced by irrelevant characteristics of the samples. For example, when asked if there is a difference in the sweetness of two samples of peach halves, a panelist may look for help in every possible way such as the following: Are the pieces of uniform size? Is there a difference in color? Is one firmer than the other? Because of this stimulus error, make all samples as uniform as possible. If unwanted differences occur between samples, mask them whenever possible.

Logical error

Closely associated with stimulus error is logical error, which can cause the panelist to assign ratings to particular characteristics because they appear to be logically associated with other characteristics. A slight yellow color in dehydrated potatoes, for example, might indicate oxidation to the panelist who could logically find a different flavor in the sample. Control this error by keeping the samples uniform and masking differences.

Leniency error

This error occurs when panelists rate products based on their feelings about the researcher, in effect ignoring product differences. Therefore, conduct tests in a controlled, professional manner.

Halo effect

Evaluating more than one factor in a sample may produce a halo effect. The panelist often forms a general impression of a product and if asked to evaluate it for odor, texture, color, and taste at the same time, the results may differ from those when each factor is rated individually. In effect, the rating of one factor influences the rating of another. For example, in meat evaluations, often panelists will rate a dry sample tougher than it would be if tenderness alone were being assessed. When resources allow, eliminate this effect by evaluating only one characteristic at a time.

Suggestion effect

Reactions of other members of the panel can influence the response of a panelist. For this reason, separate panelists from each other in individual booths. Do not permit them to talk during the testing so that a suggestion from one panelist will not influence another. Keep the testing area free from noise and distraction, and separate from the preparation area.

Positional bias (order effect)

Often panelists score the second product (of a set of products) higher or lower than expected regardless of the product because of position effect. In some tests, particularly the triangle test, a positional bias has been shown. When there are no real sample differences, panelists generally choose the middle sample as being different. Avoid this error by making either a balanced or a random presentation of samples. In a small experiment, use a balanced presentation to ensure every possible order is presented an equal number of times. In a large experiment, randomize the samples.

Contrast effect and convergence error

A contrast effect occurs between two products that are markedly different; panelists will commonly exaggerate the difference in their scores. For example, presenting a sample that is very sweet before one that is slightly sweet causes the panelist to rate the second sample lower in sweetness than it would normally be rated; the reverse is also true.

Convergence error is the opposite of contrast effect. A large difference between two (or more) products may mask small differences between other samples in the test causing scores to converge. To correct for both these errors, randomize the order of presentation of the samples for each panelist, so as to equalize these effects.

Proximity error

When a set of samples are being rated on several characteristics, panelists usually rate more similar those characteristics that follow one another (in close proximity) on the ballot sheet than those that are either farther apart or rated alone. Thus, the correlations between characteristics close together may be higher than if they were separated by other characteristics. Minimize this error either by randomizing the characteristics on the ballot sheet or by rating only one characteristic at a time.

Central tendency error

This error is characterized by panelists scoring products in the midrange of a scale to avoid extremes. It causes the treatments to appear more similar than they may actually be and is more likely to occur if panelists are unfamiliar with either the products or the test method. To minimize this error, balance or randomize the order of presentation as this effect is more noticeable for the first sample. Train panelists to familiarize them with the test method and products.

Motivation

The motivation of the panelists affects their sensory perception. An interested panelist is always more efficient and reliable and is essential for learning and good performance. Maintain the interest of each panelist by giving them reports of their results. Trained panelists are generally more motivated than those who are not trained (Ellis 1961). Help to make participants feel that the panels are an important activity by running the tests in a controlled, efficient manner.

Physical facilities

Testing area

Sensory analysis requires a special testing area that is kept constant throughout all tests and where distractions are minimized and conditions are controlled (Fig. 1). In designing an area, consider management support, location, space requirements, environmental aspects, construction, cost, and laboratory design (American Society for Testing and Materials 1986). Each type of testing, for example, discriminative and descriptive testing, focus group, or consumer panel testing, demands some modification to the design of the facilities. The testing environment should provide a quiet, comfortable environment complete with an air conditioner and source of heat to maintain 22°C. A controlled humidity of 44–45% RH may be required for testing some products (American Society for Testing and Materials 1986).

Most types of testing, excluding profile methods, require independent responses from the panelists. To accomplish this, provide the testing area with individual booths that are adjacent to a separate sample preparation and serving area. The booths may be as simple as partitions either put on a table or hinged to collapse when the laboratory bench is used for other purposes. American Society for Testing and Materials (ASTM) (1986) has recommended a booth width of between 70 and 80 cm, a depth of between 45 and 55 cm, and a countertop height of 75 cm (Fig. 2). The usual method is to construct booths along the wall that divides the testing area from the preparation area. If both countertops (preparation and testing area) are at the same height, the product can be passed through from the preparation area to the panelists. The pass-through can be either a sliding door or a bread-box style (Fig. 3). Construct all walls and booths of opaque, nonreflecting material, which is neutral in color (off-white, white, light gray), easily cleaned, and ideally divided by dividers that extend out 40 cm beyond the countertop. Panelists should not enter the preparation area, as they might gain information about the sample that may influence their responses.

The booth may be fairly simple or very elaborate depending on the funds available, the type of products being tested, and the types of tests or panels required. Many laboratories have a sink and tap built into each booth for expectoration and to provide water for rinsing. We do not recommend sinks for food and beverage testing because, if they are not

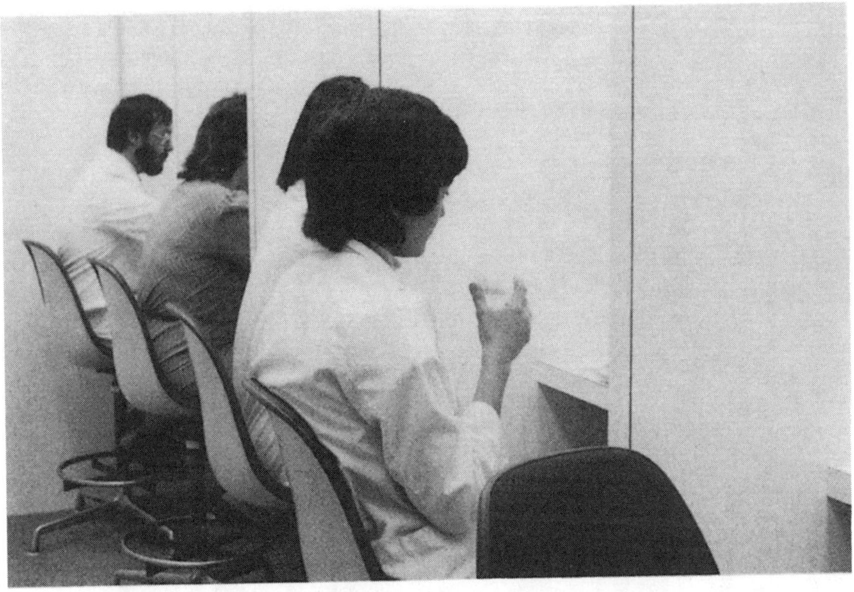

Fig. 1 Sensory analysis area. (Diversified Research Laboratories Ltd.)

properly maintained, sanitation and odor problems will result (Ellis 1961). Noise is sometimes also a problem. Sinks are often required for testing personal care products and, if so, we recommend a suction type, such as dentists use. A tap for rinsing water in the booth is undesirable because it is difficult to control the temperature of the water. It is generally advisable to pour the water well before the test so that it will be at room temperature. Water can be tap, bottled spring, or distilled. Some testing requires such items as mirrors, electrical outlets, or hot plates.

Develop some method of communication from the panelist to the researcher. In some laboratories, a switch in each booth connected to lights in the preparation area acts as the signal that a panelist is ready for the samples. Other methods are the placement of a tag with the panelist's name on it into the pass-through or a direct data entry system whereby the panelist logs in on a computer terminal.

Fig. 2 Panelist with samples and questionnaire.

Ensure that lighting is uniform and does not influence the appearance of the product being tested. Choose carefully the type of light used if color and appearance are important factors to be judged. The American Society for Testing and Materials (1986) has recommended incandescent and fluorescent lighting, which will provide a variety of intensities ranging from about 753 lux to at most 1184 lux. Install a dimmer switch to create varying intensities of light. The International Organization for Standardization (ISO) (1985) has recommended lights having a correlated color temperature of 5000/6500 K. To eliminate differences in color between samples, colored lights are sometimes used; the most common are red, green, or blue. These lights have not been particularly effective because differences in the hue or depth of color are still noticeable. Amerine et al. (1965) pointed out that it is not known what effect colored or dim lighting has on judgment. It may have less effect on experienced panelists who are accustomed to it, but inexperienced panelists have expressed a dislike for testing under colored lights. In the laboratory at the Food Research Centre, we have had panelists experience a claustrophobic feeling upon entering a room lit with colored lights. Any evaluations made under this type of stress would be questionable. Sodium lighting is another alternative for masking color differences.

Keep foreign odors and odors from the food preparation area from the testing room. Do not allow smoking in this area at any time. Panelists should also avoid cosmetics. The American Society for Testing and Materials Committee E–18 on Sensory Evaluation of Materials and Products (1986) has recommended that a slight positive pressure

Fig. 3 Presentation of sample through bread-box-style hatch.

maintained in the testing area prevents odors entering from surrounding areas. That committee also recommended the use of activated charcoal filters in the incoming air vents for an odor-free area; these filters need to be changed or reactivated at intervals.

Training area

Provide an area for training a profile or descriptive analysis panel (Fig. 4). Ideally, the area would be separate and adjacent to the preparation area, if space allows. Often it is part of the booth area. A round table, large enough for 6–12 panelists, allows discussion among the trainees. A table that has a movable centre or lazy susan for passing standards or reference samples back and forth is very useful. Ensure that the lighting and ventilation are similar to that described for the booth area and that tables, chairs, floorings, and walls are neutral in color, nonodorous, and easily cleaned. To facilitate discussion, provide a blackboard, whiteboard, or flip chart, but be aware of the odor given off by some markers. Lastly, include a clock to ensure that the panel leader does not exceed the allotted training time.

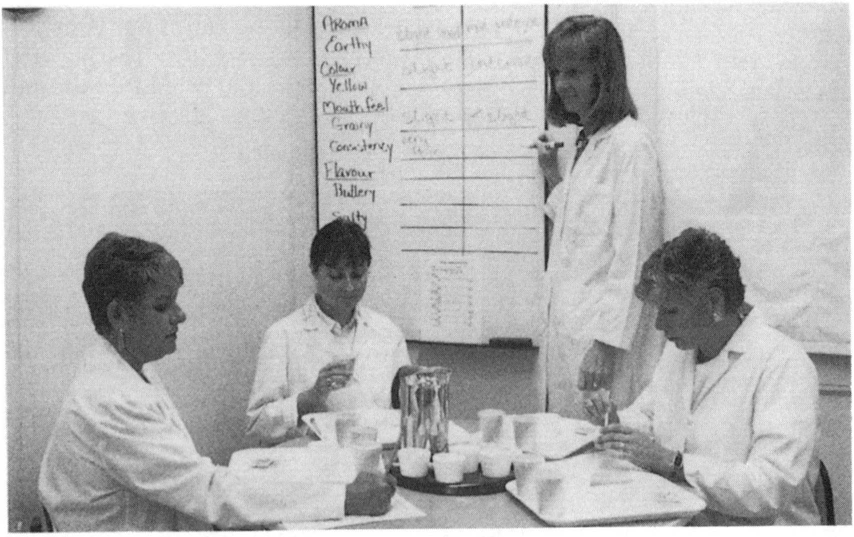

Fig. 4 Training area.

Preparation and serving area

The type of preparation area and equipment will depend on the type of products being tested. A well-equipped kitchen is a good start, with specialized equipment being added according to need. Install in the preparation area a good ventilation system to remove odors and an air conditioner to maintain a controlled temperature. We cannot overemphasize the need for sufficient counter space for serving and assembling samples for presentation and cabinet space for storage. Within the preparation area, provide refrigerated and frozen storage space to store or hold samples. Arrange for extra refrigerated and frozen storage, for larger volume storage, in another location if needed. If heating or cooking of samples is necessary, then install equipment such as electric or gas burners (the burners must be the same size) and ovens (e.g., conventional, convection, or microwave). Other equipment may include water baths, heating trays, incubators, meat cutters, balances, heat lamps, scales, mixers, bottled water dispensers, and fire extinguishers. For cleaning purposes the preparation area may include dishwashers, garbage disposers, trash compactors, in addition to the required waste baskets, sinks, and water sources. When considering equipment, address such factors as initial cost, quality, service availability, space and installation requirements, cleanability, noise level, and odor generation. When selecting materials for the construction of the preparation area, keep the colors neutral and avoid reflective materials such as glass and mirrors, which are difficult to keep clean. Vinyl coverings and tiles are suitable for walls and floors and plastic laminates for cupboards and countertops. Consider having a telephone in the preparation area for contacting panelists who are late for a panel. But we do not recommend installing a telephone in the booth area, because it is distracting if it rings.

Sample preparation

Preliminary testing is usually needed to decide on the method of preparation; thawing, preparation, and cooking time; and equipment and utensils. For difference (discriminative) testing, select the preparation method that is unlikely to mask, add to, or alter the basic characteristics of the product. For affective testing, prepare the product using a method typical of, or that represents, the way in which it is actually prepared and consumed.

Once a method has been chosen, keep it constant throughout tests on the product. The preparation method must not impart any foreign tastes or odors to the product. Samples should represent or be typical of the product or material to be tested. Except for the factor (or factors) under study, prepare all samples tested at a given time or in a given experiment and serve them using exactly the same procedures.

Panelists are influenced by irrelevant characteristics of the samples. Therefore, make every effort to prepare samples from different treatments identically to remove any irrelevant differences. To achieve this, it may be necessary to cut, dice, grind, or puree the samples to obtain uniformity of sample presentation across treatments.

When conducting affective tests, it is better to present one sample at a time rather than grind or puree the product to mask differences. Color differences are sometimes masked by colored or dimmed lights, as previously discussed. The use of colored containers such as ruby-red glasses or black-lined cups masks color differences. Dyes that impart no flavors have also been used to eliminate color differences.

Dilution and carriers

Some products, by their very nature, need either to be diluted or to be served with a carrier. For example, hot sauces, flavorings, spices, and sauces usually require dilution prior to testing. Spices can be mixed into a bland white sauce or syrup, but do consider the effect of the sauce or syrup on the flavor. For example, several researchers have shown that hydrocolloid gels have flavor-masking effects (Marshall and Vaisey 1972; Pangborn and Szczesniak 1974).

Some products, such as whipped topping, cheese spread, and ketchup, do not require carriers (Kroll and Pilgrim 1961). In their study, they found that discrimination was better without a carrier in some cases. The use of a carrier, such as crackers for jam or frankfurters for ketchup, adds to the cost and effort, and it is often difficult to select an appropriate carrier. Carriers are also a source of experimental error because either the proportion of product to carrier may not be constant, or the carrier may not be of consistent quality.

The nature of a product may sometimes require that it be tested with a carrier during some of the testing. Test pie filling with pastry to see how the two interact, especially in terms of texture. Test icing on cake not only to examine flavor combinations, but also to see how the icing handles.

Serving temperature

The temperature at which samples are served can cause many problems. For affective tests, serve the samples at the temperature at which they are normally consumed.

However, in discriminative or descriptive tests, modify the temperature to account for the fact that taste acuity or perception is considered to be greatest at a temperature between 20° and 40°C. The temperature throughout the experiment must remain constant for the results to be comparable. To achieve this, use warming ovens with a controlled thermostat, water baths, styrofoam cups, heated sand, doubleboilers, or wrap in foil. If the samples are to be held for any length of time, take care that they do not dry out or change in quality during holding. For example, do not cut meat samples until ready to serve; otherwise they will dry out, which will affect the sensory rating.

Utensils and containers

Serving utensils should impart neither taste nor odor to the product. Use identical containers for each sample so that no bias will be introduced from this source. Unless differences in color are being masked, it is wise to use colorless or white containers. Consider factors such as ease of coding, type of product, and serving temperature when you select containers. Disposable dishes made from hard plastic, unwaxed paper, or styrofoam are convenient when large numbers are to be served. Determine beforehand that no taste is transferred to the product. Glass is an excellent vessel; however, it requires time spent in washing and rinsing to ensure that no flavor or odor is left from the products or soap used. Breakage of glass is also a problem.

Quantity of sample

The amount and size of sample given to each panelist is often limited by the quantity of experimental material available. The sample, even if small, should be representative of the product. Even when using a small amount of sample, each panelist must have enough to assess the product and retaste, if necessary. Make the amount of sample presented constant throughout the testing. When using a reference sample, we recommend that you present twice as much sample as the experimental sample to allow panelists to keep referring back to the reference.

The Committee E–18 of the American Society for Testing and Materials (1968) has recommended that, in discriminative tests, each panelist should receive at least 16 mL of a liquid and 28 g of a solid, and that the amount should be doubled for preference tests. In some tests, panelists have been asked to consume a normal serving size of a product. For example, during acceptability studies on flavored milk, panelists were asked to drink 190 mL of the product before rating the acceptability. Such a product might be very pleasing when you drink 30 mL, but if you drink 190 mL you might find it too sweet or satiating.

Number of samples

With input from panelists, in preliminary work, determine the number of samples that can be effectively evaluated in one session. Consider not only the type of product, the number of characteristics to be evaluated, the type of test, and the experience of the panel, but also the motivation of the panelists. The number of samples that can be presented in a given session is a function of both sensory and mental fatigue in the panelist (Meilgaard et al. 1987*a*).

Reference samples

Inclusion of a known (or marked) reference sample (or samples) may assist panelists in their responses and may decrease the variation in judgment. For example, if you cannot present all the samples at once, the use of a reference sample may be helpful. The reference sample must be the same each day of evaluation throughout the experiment and can be anchored to the scales. A reference sample included as a coded sample provides a check on a panelist's consistency in evaluations.

Coding

The code assigned to the samples should give the panelists no hint of the identity of the treatments, and the code itself should not introduce any bias. We recommend three-digit random numbers obtained from tables of random numbers for coding the samples. A table of random numbers is presented in Statistical Chart 1 (Appendix). Enter anywhere on the table and, by moving either horizontally or vertically, select three consecutive numbers. If replication is being done, the panelists are usually the same, therefore, a new set of codes is required for each replicate. Because some marking pens can leave an odor, use a wax pencil particularly for odor testing to mark containers with the number. Computer-generated labels can also be used.

Order of presentation

Many psychological and physiological effects, which we have discussed earlier, make it necessary to have an order of presentation that is either balanced or random. With a small number of samples and panelists, a balanced order makes it possible for every order to occur an equal number of times. In larger experiments, randomize the order using tables of random numbers or by computer using statistical software. To obtain a random order presentation, first assign a number to each treatment, e.g., sample A = 1, sample B = 2..., etc. Then, randomly enter the random number table (Statistical Chart 1) and determine the order of presentation as you come across each treatment number (horizontally or vertically).

Rinsing

Provide panelists with an agent for oral rinsing between samples. Many researchers prefer taste-neutral water at room temperature, but, when fatty foods are being tested, warm water, warm tea, lemon water, or a slice of apple or Japanese pear is a more effective cleansing agent. Unsalted crackers, celery, and bread have all been used for removing residual flavors from the mouth. Some researchers insist that the panelists rinse between each sample and others allow them to rinse or not according to personal preference. Whatever the case, each panelist should follow the same procedure consistently after each sample. In some cases, you may need to control the time between samples to prevent carryover from one sample to another, but panelists are usually allowed to work at their own speed.

Information about sample

Give the panelists as little information as possible about the test to avoid influencing results. If they are given information about samples, panelists will have some pre-conceived impression of what to expect. Because this expectation error exists, do not include on the panel persons who are directly involved with the experiment.

Time of day

The time of day that tests are conducted influences the results (Amerine et al. 1965). Although time cannot always be controlled if the number of tests is large, late morning and mid afternoon are generally the best times for testing.

Consider the type of product being sampled. Too early in the morning or afternoon is objectionable to some panelists, especially so if the foods served are hot, spicy ones. If it is too late in the day, some panelists may lack motivation. Avoid mealtimes. Set the schedule taking into consideration other testing to be done.

Selection and training of panelists

The panel is the analytical instrument in sensory analysis. The value of this tool depends on the objectivity, precision, and reproducibility of the judgment of the panel members. Before a panel can be used with confidence, the ability of the panelists to reproduce judgments must be determined. Panelists for descriptive testing need to be carefully selected and trained.

Panelists for laboratory testing are usually office, plant, or research staff. It should be regarded as part of a normal work routine for personnel to participate on a panel; they should be expected to evaluate all products. However, do not ask any person to evaluate food to which they object or may have an allergic reaction. Management support and full cooperation of the supervisors of persons who serve as panelists are necessary. A small, highly trained panel will give more precise and consistent results than a large, untrained panel.

Selection criteria

Selection is essential to develop an effective descriptive sensory panel. Select panelists on the basis of certain personal characteristics and potential capability in performing specific sensory tasks. Include in the selection criteria health, interest, availability, punctuality, and verbal skills.

Health

Persons who serve as panelists should be in good health and should excuse themselves when suffering from conditions that might interfere with normal functions of taste and smell. For example, colds, allergies, medications, and pregnancy often affect taste and smell sensitivities. We recommend that panelists refrain from smoking, chewing gum, eating, or drinking for at least 30 min before testing. Keep records of panelists' allergies, likes, and dislikes.

Interest

Emotional factors, interest, and motivation appear to be more important than the age or sex of panelists. Motivation affects their response. Interest is essential for learning and good performance, but often it is difficult to maintain a panelist's interest. One solution could be to provide each with the test results. You can help panelists to feel that panels are an important activity and that their contribution is important by running the tests in a controlled, efficient manner.

Availability and punctuality

Availability of the panelist during training and testing is essential. Persons who travel frequently and some production personnel are unsuitable to serve on panels. Punctuality is essential not only to avoid wasting people's time but also to avoid loss of integrity in sample and experimental design. Encourage punctuality by providing advance notice of all tests, i.e., a test schedule, regularly scheduled test sessions, and a personal reminder or telephone call shortly before test sessions.

Verbal skills

The degree of verbal skills that is required of the panelist depends on the test methodology. Descriptive tests generally require good verbal communication skills because panelists are expected to define and describe various characteristics of products.

Selection and screening

Amerine et al. (1965) believed that by selecting and training panelists with consistent, discriminative abilities, panels could be small and efficient. Threshold tests are not useful in selecting panelists because sensitivity to the primary tastes may not be related to the ability to detect differences in food. A more realistic approach is to select panelists on their ability to detect differences in the food to be evaluated. Ideally you should screen two to three times as many people as you will need, using the product class that will be tested. Prepare the samples so the variation you obtain is similar to that which the panel will find in the actual experiment. When possible, the test methods used for screening should be similar to the actual ones to be used during testing. Ensure that each person clearly understands each test method, score sheet, and evaluation technique. Rank each person according to their ability to differentiate among (for discriminative tests) or describe (for descriptive and profile tests) the samples prepared. The selected panelists should have inherent sensitivity to the characteristic being evaluated. Repeat the tests to get a measure of reproducibility. Select a new panel for each product. Persons who discriminate well on some products do not necessarily discriminate well on others.

Discriminative tests include triangle, duo–trio, two-out-of-five, paired comparison, and ranking test methods. Screening for a discriminative panel requires either that you present a series of triangle tests to each panelist and calculate the percentage of correct identification of odd or different samples (American Society for Testing and Materials 1981); or that you perform a sequential screening analysis (American Meat Science Association 1978).

Descriptive analysis includes flavor profile, texture profile, quantitative descriptive analysis, and attribute rating methods. For the purpose of screening for flavor analysis, for which essentials are flavor memory and ability to deal logically with flavor perceptions, American Society for Testing and Materials (1981) recommended certain tests

- to determine the panelist's ability to differentiate the basic tastes at above-threshold levels
- to determine the panelist's aptitude for identifying and describing 20 different odorants
- to test the panelist's ability to rank basic taste samples in order of increasing concentrations.

If texture analysis is required, test panelists for their ability to rank samples within various texture scales, e.g., hardness, viscosity, and geometrical characteristics in increasing order of attribute intensity (American Meat Science Association 1978).

Training

Training improves an individual's sensitivity and memory to provide precise, consistent, and standardized sensory measurements that can be reproduced. For panelists to make objective decisions, they must be trained to disregard their personal preferences. Training involves the development of a vocabulary of descriptive terms. Each panelist must detect, recognize, and agree upon the exact connotation of each descriptive term. The use of specially prepared reference standards or competitor's products that demonstrate variation in specific descriptive terms can help panelists during training sessions become more consistent in their judgments. Discuss the evaluation techniques for odor, appearance, flavor, and texture and agree upon a common procedure. Panelists must also become familiar with the test method. Training time (from weeks to months) is a function of the product, the test procedure, and the capability of the panelists.

Experimental design

In planning experiments consider carefully the hypotheses to be tested, size of experiment, replication, blocking, randomization, and statistical tests.

Samples

Samples are taken to learn about the population being studied. The population is the total of all possible observations of the same product from which a sample is drawn. Characteristics will vary from sample to sample within a population. Therefore, when decisions about a population are based on samples, it is necessary to make allowance for the role of chance. Ensure that the samples fully represent the population from which they are drawn.

Hypotheses to test

Before selecting the test, the sensory analyst must establish the objective of the study, which is then stated in the form of a null hypothesis. For example, when testing two products having different levels of sweetener to see if the difference in sweetness is detectable, the null hypotheses might be as follows: "There is no difference in sweetness between these products."

Based on the results of the statistical analysis of the experimental data, we either accept or reject the null hypothesis. Associated with the decision to accept or reject the null hypothesis are two types of error. A Type I error occurs when the null hypothesis is rejected when it is true; that is, saying there is a difference when in fact there is none. A Type II error occurs when the null hypothesis is accepted when in fact it is false; in other words, saying there is no difference when there really is one.

The probability of making a Type I error is the level of significance (α). Usually the level of significance is set at 0.05 (5%) or 0.01 (1%). The 0.05 level of significance means there is 1 chance out of 20 of saying there is a difference when there is no difference. A result is considered to be significant if the probability (P) is 0.05 or less. The probability of making a Type II error is β.

When working in research, statistical tests are usually set up to find differences. Therefore, Type I errors should concern us, namely, that we reject the null hypothesis when we should not reject it. However, often in sensory analysis, for quality assurance purposes, new products are tested against standard products to ensure that they do not differ. In this instance, the Type II error should concern us, namely, that we say there is no difference when there really is a difference. If the latter occurs, consumers would notice the change when the new product is marketed and would possibly stop purchasing it. Therefore, we must minimize β by using acute, reliable panelists and by increasing the sample size (Larmond 1981). We refer the reader to O'Mahony (1986) for further discussion.

Size of experiment

The number of judgments collected will influence the statistical significance of the results. If too few judgments are obtained, large differences are required for statistical significance, whereas with a large number of judgments, statistical significance may result when differences are very small. Although statistical significance is important when reporting results, the size of the differences is also important. For example, a difference of 0.3 cm on a 15-cm line is unlikely to be meaningful (Larmond 1981). If prior information on variability is available, an appropriate sample size can be estimated, which will result in those meaningful differences that do exist being statistically significant; see, for example, Steel and Torrie (1980) or Gacula and Singh (1984). If no information is available on variability, a preliminary study could provide this.

Replication is necessary to provide an estimate of experimental error. Ideally, the sensory analyst should ensure that the replicates are independent units that represent the population. For example, if different varieties of applesauce were being evaluated, each panelist should receive applesauce from different tins (independent), rather than from the same tin. In some cases this design is not practical. For example, if roasts from different breeds of animals were being evaluated, several panelists might evaluate samples from the same roast (subsamples). The experimental unit in this case is the roast, not each panelist's sample, and we should compare the effect of breed to the among-roast variability. For further discussion of this aspect of design, see chapter 12 of Meilgaard et al. (1987*b*).

Blocking

It is generally possible to increase the power (ability to detect real differences) of an experiment by removing known sources of variability from the estimate of error. For example, panelists often use different parts of a scale when making judgments. We can remove this variability from the error by pairing or blocking observations. That is, each panelist evaluates all treatments. Make samples for each panelist as homogeneous as possible so that comparisons will be as precise as possible.

Treatments compared at the same time by the same panelist (in one block) are generally more precisely compared than those judged by panelists on different occasions. This may cause severe design restrictions. Differences in large numbers of treatments will be more difficult to estimate precisely than differences in a small number of treatments that a panelist can compare in one sitting. Further, because of the number of treatments being considered or some particular characteristic of the product, such as a lingering aftertaste, it is not always possible to evaluate all treatments in one sitting. In these instances, an incomplete block design is used. Choose the most efficient design (smallest variance) possible. More replications are required to obtain the same efficiency as with a complete block design. When all treatments cannot be compared in one sitting, include controls, if possible, to improve comparability across sessions (Gacula 1978). For further details on complex designs, we suggest that you consult a statistician. This subject is discussed in more detail in Cochran and Cox (1957), Moskowitz (1988*b*), Meilgaard et al. (1987*b*), and Gacula and Singh (1984).

Randomization

Proper randomization is essential for valid, unbiased results. To guard against a treatment unknowingly being favored or disfavored throughout an experiment and to ensure an unbiased estimate of error, it is important to randomize. Items to be randomized include, for example, the order of presentation, the oven used in preparation, and the assignment of material to a treatment. The random number tables (presented as Statistical Chart 1) will assist in establishing a random ordering at every step of the experiment.

Statistical tests

In general, discriminative tests are analyzed by comparing test statistics with chart values; ranking test results are analyzed by calculation of a Friedman statistic; structured, unstructured, and ratio scale tests with two samples are analyzed by a *t*-test; and tests with more than two samples are analyzed by an *F*-test. The following analyses are described in the text:

- triangle test (page 20)
- duo–trio test (page 22)
- two-out-of-five test (page 23)
- paired comparison test (page 25)
- Friedman for ranked data (page 26)
- *t*-test (page 32)
- analysis of variance (page 36)
- analysis of variance with interactions (page 42)
- analysis of variance of logarithmically transformed data (page 48).

When there are more than two treatments, the sensory analyst may wish to know more than whether there is a significant difference among the samples. Use of a multiple comparison test can determine which pairs of means are significantly different. Multiple comparison tests are not always appropriate. Available multiple comparison tests include Scheffé, Tukey, Newman-Keuls, Duncan's multiple range, and Fisher's LSD. A more powerful test (e.g., Fisher's LSD) has a smaller least significant difference

(LSD) and therefore a greater likelihood of finding a difference. However, there is also the risk of finding such a difference regardless of whether or not one exists. On the other hand, a more conservative test (e.g., Tukey's) has a larger LSD and therefore is less likely to find a difference between the means. A Dunnett test is a multiple comparison test designed for a special application; it is used only when all the means are compared to one of the means, for example, a control. In some cases, such as increasing or decreasing concentrations of a factor, it is more appropriate to perform a linear regression on the samples. This will determine the degree of association between two sets of data, namely, sensory results and the factor under study. The reader can find details of each of these tests in *Sensory evaluation of food—Statistical methods and procedures* (O'Mahony 1986).

The following comparison tests are described in the text:

- multiple comparison test (Tukey's) for scaling (page 38)
- multiple comparison test for ranks (page 28)
- comparison of control (Dunnett's) to other scores (page 45)
- estimation of linear regression (page 49).

Sensory analysis test methods

Several different sensory analysis methods are now available. The researcher should be thoroughly familiar with the advantages and disadvantages of each method. Choose the most practical and efficient method for each situation. No one method can be used universally. The researcher must precisely define the purpose of the test and the information sought from it.

The three fundamental types of sensory tests are discriminative tests, descriptive tests, and affective tests. We use discriminative tests to determine whether a difference exists between samples. We use descriptive tests to determine the nature and intensity of the differences. Affective tests are based on either a measure of preference (or acceptance) or a measure from which we can determine relative preference. The personal feelings of panelists toward the product directs their response. In this publication, we describe several commonly used experimental methods with examples of the questionnaires, their application, and statistical analysis.

Discriminative tests

Sensory analysis is often conducted to determine whether or not a difference exists among samples. Testing for "sameness" is referred to as "similarity" testing. Typically in quality control this type of testing predominates. It is necessary then to minimize β (Type II error). Testing to find a difference is referred to as "discriminative" (difference) testing. The α value (Type I error) is therefore minimized. We direct the reader to Meilgaard et al. (1987*a*,*b*) for a more detailed explanation of similarity and difference testing. The latter is a common situation in quality maintenance, cost reduction, selection of new sources of supply, and storage stability studies. Several different sensory methods allow us to determine differences. The difference tests included in this publication are triangle test, duo–trio test, two-out-of-five test, paired comparison test, and ranking test. We discuss the advantages, disadvantages, and special features of each test.

Triangle test

The results of a triangle test indicate whether or not a detectable difference exists between two samples. Higher levels of significance do not indicate that the difference is greater but that there is a greater probability of a real difference.

The panelist receives three coded samples, is told that two of the samples are the same and one is different, and is asked to identify the odd sample. This method is useful in quality control work to determine if samples from different production lots are different. It is also used to determine if ingredient substitution or some other change in manufacturing results in a detectable difference in the product. The triangle test is often used in selecting panelists.

Because the panelist is looking for the odd sample, the samples should differ only in the variable being studied. Mask all other differences. Therefore, application of the triangle test is limited to products that are homogeneous.

Analysis of the results of triangle tests is based on comparing the number of correct identifications actually received with the number you would expect to get by chance alone if there were no difference between the samples. In the triangle test we would expect the odd sample to be selected by chance one-third of the time.

To determine the probability that the different sample was correctly identified by chance alone (no detectable difference), use Statistical Chart 2 (Appendix).

With the triangle test, neither are the size and direction of the difference between samples determined nor is there any indication of the characteristic responsible for the difference.

There are six possible ways in which the samples in a triangle test can be presented:

ABB	BBA	AAB
BAB	ABA	BAA

Indicate the order in which each panelist should taste the samples by putting the code numbers in the appropriate order on the score sheet.

In most cases, each sample is used as the duplicate for half the tests and as the different sample for the other half. In some cases it has been found that more correct identifications are received when the duplicate samples are the normal or control samples. A sample of the questionnaire and an example of the triangle test follow.

QUESTIONNAIRE FOR TRIANGLE TEST

PRODUCT: Canned tomatoes

NAME ____________________ DATE ________________

Two of these three samples are identical, the third is different. Taste the samples in the order indicated and identify the odd or different sample.

Identify the odd or different sample.

Code	*Check odd or different sample*
314	________________
628	________________
542	________________

Comments:

Example A triangle test was used to determine if there was a detectable difference between canned tomatoes processed under two different sets of conditions.

Samples were served in coded dishes to 36 panelists. Each panelist received three coded samples: 18 panelists tested two samples from treatment A and one from treatment B; the other 18 panelists received one sample from treatment A and two from treatment B. Because of the nature of the presentation, it was necessary to assign two code numbers to each treatment. The results are shown in Table 1.

Treatment	*Code*	*Number of samples required*
A	314	18
	542	36
B	628	36
	149	18

Table 1 Triangle test on canned tomatoes processed by treatment A or treatment B

Code (treatment)	Odd sample chosen Correct	Incorrect	Total
314(A) 628(B) 542(A)	9	9	18
542(A) 628(B) 149(B)	12	6	18
Total	21	15	36

The odd sample was correctly identified by 21 panelists. According to the Statistical Chart 2 (Appendix), 21 correct judgments out of 36 in a triangle test indicate a probability of 0.002, which is less than the critical value of $P = 0.05$. Thus the probability of getting these results by chance is 2 in 1000. We conclude that a detectable difference existed between the samples.

Duo–trio test

In the duo–trio test, three samples are presented to the panelist; one is labeled R (reference) and the other two are coded. One coded sample is identical with R and the other is different. The panelist is asked to identify the odd sample. This test is similar to the triangle test, except that, one of the duplicate samples is identified as the reference (R).

The duo–trio test has the same applications as the triangle test but is less efficient because the probability of selecting the correct sample by chance is 50%. Often panelists find this test easier than the triangle test. This test is often used instead of the triangle test when evaluating samples that have a strong flavor because fewer comparisons are required. The two coded samples do not have to be compared to each other.

For the duo–trio test either one of the samples can be selected for use as the reference throughout the whole test or the two samples can be used alternately as the reference. The quantity of the R sample should be about double that of the coded samples.

Determine the likelihood that the number of correct identifications was obtained by chance alone (no detectable difference). The duo–trio is always a one-tailed test (see "Paired comparison test"). Use Statistical Chart 4 (Appendix) to determine this probability. A sample of the questionnaire and an example of the duo–trio test follow.

QUESTIONNAIRE FOR DUO–TRIO TEST

PRODUCT: Cheddar cheese

NAME ______________________ DATE ______________________

On your tray you have a marked control sample (R) and two coded samples. One sample is identical with R and the other is different.

Which of the coded samples differs from R?

Code	*Check odd or different sample*
432	______________________
701	______________________

Comments:

Example A duo–trio test was used to determine if methional could be detected when added to cheddar cheese at 0.250 ppm. The duo–trio test was used in preference to the triangle test because less tasting is required to form a judgment. This fact is important when panelists taste a substance with a lingering aftertaste, such as methional.

The test was performed using 16 panelists. The panelists were presented with a tray on which were a coded sample containing 0.250 ppm methional and two control samples, one R and one coded. The order of presentation of the two coded samples was randomized. A total of 16 judgments were made. The results showed that, of the 16 judgments made, 14 correct identifications were obtained.

Consult Statistical Chart 4 (Appendix) for 16 panelists in a two-sample test. This chart shows that 14 correct judgments has a probability of 0.002, which is less than the critical value of $P = 0.05$. We conclude that methional added to cheddar cheese was detectable at the 0.250 ppm level.

As with the triangle test, the duo–trio test can establish if a detectable difference exists between samples. However, it does not indicate the size of the difference or whether the panelists' identifications of the odd sample were based on the same characteristic.

Two-out-of-five test

The panelist receives five coded samples, is told that two of the samples belong in one set and three to another, and is asked to identify the two samples that belong together.

This method has applications similar to the triangle test. It is statistically more efficient than the triangle test because the probability of guessing the right answer in the two-out-of-five test is 1 in 10, compared to 1 in 3 for the triangle test. However, this test is strongly affected by sensory fatigue. It is recommended for visual, auditory, and tactile testing rather than flavor and odor.

Analysis of the results of the two-out-of-five test is based on the probability that if there were no detectable difference, the correct answer would be given one-tenth of the time. A table for rapid analysis of two-out-of-five data is presented in Statistical Chart 5 (Appendix).

The results of the two-out-of-five test indicate if there is a detectable difference between two sets of samples. No specific characteristic can be identified as responsible for the difference.

Twenty presentation orders are possible in the two-out-of-five test:

AAABB	ABABA	BBBAA	BABAB
AABAB	BAABA	BBABA	ABBAB
ABAAB	ABBAA	BABBA	BAABB
BAAAB	BABAA	ABBBA	ABABB
AABBA	BBAAA	BBAAB	AABBB

If 20 panelists are not used, select at random from the possible combinations using three As for half and three Bs for half. A sample of the questionnaire and an example of the two-out-of-five test follow.

QUESTIONNAIRE FOR TWO-OUT-OF-FIVE TEST

PRODUCT: Wieners

NAME ________________________ DATE ________________

Two of these five samples belong to one set; the other three belong to another set. Examine the color of the samples in the order listed below.

Identify the *two* samples which belong together by placing an × after the code numbers.

Code

846 ________
165 ________
591 ________
497 ________
784 ________

Comments:

Example A two-out-of-five test was used to determine if a different curing agent changed the color of wieners. Coded samples were presented to 20 panelists who were asked to examine them visually. The method of evaluation specified depends on the purpose of the test, e.g., listen to the crunch when chewed, or feel the roughness of the sample.

The number of correct responses was counted:

Total responses:	20
Correct responses:	7

According to Statistical Chart 5 (Appendix), 7 correct responses from 20 panelists indicates a difference significant at $P = 0.002$. We conclude that a significant difference in color existed between the samples.

Paired comparison test

In this test, the panelist receives a pair of coded samples and is asked to compare for the intensity of some particular characteristic. The panelist is asked to indicate which sample has greater intensity of the characteristic being studied.

The paired comparison test is used to determine if two samples differ in a particular characteristic. The difference tests, except two-out-of-five, presented up to now did not specify any particular characteristic; panelists based their answers on any detectable difference. The paired comparison test can be used for quality control, to determine if a change in production has resulted in a detectable difference. It can also be used in selecting panelists. The probability of a panelist selecting a sample by chance is 50% in the paired comparison test.

In a paired comparison test, one sample is usually expected to have greater intensity of the characteristic being evaluated. This is a directional difference test, so we use a one-tailed test (Statistical Chart 4 in Appendixes). However, when there is no expectation about the result, use a two-tailed test (Statistical Chart 3 in Appendixes).

Paired comparison tests give no indication of the size of the difference between the two samples but determine whether there is a detectable difference in a particular characteristic and the direction of the difference. A sample questionnaire and example for a paired comparison test follow.

QUESTIONNAIRE FOR PAIRED COMPARISON TEST

PRODUCT: Canned peaches

NAME ______________________ DATE ________________

Evaluate the sweetness of these two samples of canned peaches. Taste the sample on the left first.

Indicate which sample is sweeter by circling the number.

581 *716*

Comments:

Example To determine if there was a difference in sweetness between peaches canned in liquid sucrose at 45° Brix or in a 52% invert syrup at 45° Brix, a paired comparison test was used.

Peaches from each treatment were served in coded dishes to 20 panelists. Ten panelists were asked to taste one sample first; 10 were asked to taste it second. Twelve panelists chose the 52% invert syrup sample as sweeter. Invert syrup is generally considered to be sweeter than sucrose at the same concentrations; thus a one-tailed test is used. According to Statistical Chart 4 (Appendix), in a paired comparison test the probability is 0.252, which is greater than the critical value of $P = 0.05$. Therefore, we conclude that no detectable difference existed in sweetness between the two treatments.

Ranking test (Friedman)

The ranking test is an extension of the paired comparison test. The panelist receives three or more coded samples and is asked to rank them for the intensity of some specific characteristic.

The ranking method is rapid and allows several samples to be tested at once. It is generally used to screen one or two samples from a group rather than to test all samples thoroughly. The results of a ranking test can be checked for significance using the Friedman test for ranked data. No indication of the size of the differences between samples is obtained. Samples ranked one after the other may differ greatly or only slightly but they are still separated by a single rank unit. Because samples are evaluated only in relation to each other, results from one set of ranks cannot be compared with the results from another set of ranks unless both contain the same samples. If ties are permitted, an alternative test statistic must be used, the reader is directed to Meilgaard et al. (1987*a*). A sample questionnaire and an example of the ranking test follow.

QUESTIONNAIRE FOR RANKING TEST

PRODUCT: Fruit drink

NAME ______________________ DATE ______________________

Rank these samples for sweetness. The sweetest sample is ranked first, the second sweetest sample is ranked second, the third sweetest sample is ranked third, and the least sweet sample is ranked fourth. Test the samples in the order indicated.

Place the code numbers on the appropriate lines.

212 *336* *471* *649*

Ranking: Most sweet 1. ________
2. ________
3. ________
Least sweet 4. ________

Comments:

Example A ranking test was used to compare the sweetness of a fruit drink made using four different sweetening agents. Eight panelists ranked the four drinks using the score sheet above. The ranks given the samples by the panelists are shown in Table 2.

Table 2 Rank scores of four sweetening agents in fruit drink

Panelists	Treatments A (212)	B (336)	C (471)	D (649)
1	4	2	1	3
2	4	3	1	2
3	3	1	2	4
4	3	2	1	4
5	4	1	2	3
6	4	3	1	2
7	4	2	1	3
8	4	1	2	3
Rank sum	30	15	11	24

Friedman test The results are analyzed using the Friedman test for ranked data:

First calculate test statistic, T.

T = {12/[number of panelists × number of treatments × (number of treatments + 1)]} × (sum of the squares of the rank sum of each treatment) – 3 (number of panelists)(number of treatments + 1)

$= [12/(8 \times 4 \times 5)] \times (30^2 + 15^2 + 11^2 + 24^2) - 3(8 \times 5)$

$= (12/160) \times 1822 - 120$

$= 136.65 - 120$

$= 16.65.$

When the number of judgments are sufficiently large, use Statistical Chart 6 (Appendix) to find the value of chi-square, χ^2, with 3 degrees of freedom for $\alpha = 0.05$. The appropriate degrees of freedom are determined as one less than the number of samples. The value is 7.81. Exact probabilities are available in O'Mahony (1986) for small numbers of treatments and panelists.

The calculated value of T is 16.65. This is greater than the critical value of 7.81, so we conclude that a significant difference in sweetness existed among the samples ($P \leq 0.05$).

Treatments that are significantly different from one another can be determined using a test given in Hollander and Wolfe (1973) and illustrated here. This is used when a multiple-range test is appropriate (see *Statistical tests*). For less than eight panelists, exact values are given in Newell and MacFarlane (1987).

The least significant difference is determined using Statistical Chart 7 (Appendix). In this case, there were four treatments. The values listed under infinite degrees of freedom for error are always used for this rank test. The value is 3.63. The least significant difference is calculated as follows:

$$\text{LSD rank} = 3.63\sqrt{[\text{No. panelists} \times \text{No. treatments} \times (\text{No. treatments} + 1)]/12}$$

$$= 3.63\sqrt{(8 \times 4 \times 5)/12}$$

$$= 3.63\sqrt{13.33}$$

$$= 13.3.$$

Any two samples where rank sums differ by more than 13.3 are significantly different.

Compare C with each of the other samples:

$$A - C = 30 - 11 = 19 > 13.3$$
$$D - C = 24 - 11 = 13 < 13.3$$
$$B - C = 15 - 11 = 4 < 13.3$$

therefore C is significantly sweeter than A.

Compare B with the others:

$$A - B = 30 - 15 = 15 > 13.3$$
$$D - B = 24 - 15 = 9 < 13.3$$

therefore B is significantly sweeter than A.

Then compare D and A:

$$A - D = 30 - 24 = 6 < 13.3$$

These results can be shown by using letters to indicate differences:

	C	B	D	A
Rank sum	11a	15a	24ab	30b
Average rank	1.4	1.9	3	3.8

Any two rank sums not followed by the same letter are significantly different ($P \leq 0.05$). C and B are significantly sweeter than A.

A test for all treatments versus control is also available (Newell and MacFarlane 1987) and should be used when appropriate.

Descriptive tests

The sensory analyst is often interested in obtaining more information than just "Is there a difference?" Descriptive analysis can be used to identify sensory characteristics that are important in a product and give information on the degree or intensity of those characteristics. This provides an actual sensory description of test products. Descriptive information can help in identifying ingredient or process variables responsible for specific sensory characteristics of a product. This information can be used to develop new products, to improve products or processes, and to provide quality control. This section is divided into two parts. The first part describes three different scales that can be used to measure the perceived quantity of specified sensory characteristics. The second part presents three descriptive analysis methods used to obtain a descriptive profile of a product and an example of a panel procedure used at the Food Research Centre.

Scaling methods

Structured scaling

Structured or category scales provide panelists with an actual scale showing several degrees of intensity or magnitude of a perceived sensory characteristic. The intensities or response categories of the sensory attribute can be labeled with numbers, words, or a combination of the two. Usually the number of response categories used ranges from 6 to 10. A number or words (or both) can be assigned to each response category, just to the extremes, or to any combination of the two. Panelists can evaluate one to several sensory characteristics at a time for one or more products.

Descriptive words on the scale must be carefully selected and the panelists trained so that they agree on the meaning of the terms. Objective terms, such as "very hard," rather than preference terms, such as "much too hard," must be used. Panelists are not typical consumers and their likes and dislikes are not solicited. The scale must also be labeled, indicating an increase or decrease in the intensity of the characteristic being measured. A scale running from extremely sweet to extremely sour is incorrect, because sweet and sour are not opposites. A product can be both sweet and sour at the same time. Two scales must be used, one for sweetness (no sweetness to extremely sweet) and one for sourness (no sourness to extremely sour). Opposites are used with a bipolar scale (e.g., hard to soft).

Certain problems are inherent with structured scales, of which the researcher should be aware. The psychological distance or sensory interval between two descriptors might not always be equal. For example, a scale used to measure perceived sweetness of a beverage might include the terms "extremely sweet, very sweet, moderately sweet, slightly sweet, trace of sweetness, not sweet." The psychological distance between "extremely sweet" and "very sweet" is not necessarily the same as between "trace of sweetness" and "not sweet." However, the numerical distance in each case is one. Also, panelists usually avoid the extreme points on the scale, believing that another sample might have an even higher or lower intensity of the sensory characteristic (central tendency error). A nine-point scale, for example, is used as if it were a seven-point scale. To use structured scales effectively, all the panelists must be judging the same characteristic; their use is not a problem when a simple characteristic like sweetness is involved. When a complex characteristic is judged (for example, the "texture" of cheese) it must be characterized into individual components, such as hardness, cohesiveness, fat properties, and so on, and each one evaluated. All panelists may not have the same concept and therefore need training.

Including standards at various points on the scale helps to minimize panel variability. These standards act as anchors in counteracting the tendency of scales to drift in meaning with time. This instability is a marked disadvantage when structured scales are used in storage stability studies over an extended time. A sample of the questionnaire and an example of a structured scale follow.

QUESTIONNAIRE FOR STRUCTURED SCALE

PRODUCT: Cheddar cheese

NAME ______________________ DATE ______________________

Evaluate these cheese samples for bitterness.
Indicate the amount of bitterness in each sample on the scales below.

590		*172*	
____	not bitter	____	not bitter
____		____	
____	slightly bitter	____	slightly bitter
____		____	
____	moderately bitter	____	moderately bitter
____		____	
____	very bitter	____	very bitter
____		____	
____	extremely bitter	____	extremely bitter

Comments:

Example The structured scale was used to determine any difference in bitterness in cheddar cheese made using two different milk-coagulating enzymes. Samples of cheese from each treatment were coded and presented to 20 panelists for evaluation. The order of presentation was balanced so that each order (AB, BA) was used 10 times. Numerical scores were assigned to the scale with "not bitter" equal to 1 up to "extremely bitter" equal to 9. The scores are tabulated (Table 3).

Table 3 Scores using a structured scale to measure bitterness in cheese samples

Panelist	Treatments A (590)	B (172)	Difference (A – B)
1	4	4	0
2	5	4	1
3	5	2	3
4	5	2	3
5	2	4	–2
6	5	4	1
7	6	3	3
8	6	3	3
9	7	3	4
10	4	1	3
11	6	2	4
12	3	2	1
13	6	4	2
14	5	5	0
15	2	3	–1
16	3	3	0
17	6	2	4
18	4	1	3
19	4	4	0
20	5	5	0
Total	93	61	32
Mean	4.7	3.1	1.6

In this example, each panelist compared two samples. A paired *t*-test is used to analyze the data. This test assumes that the intervals between categories are equal, which may not be the case. Therefore the results should be considered approximate. Alternatively the data could be examined using a paired comparison test or a Wilcoxin signed rank test as described by O'Mahony (1986).

Analysis using t*-test*

Calculate $\overline{d}$ (average difference):

$$\overline{d} = \text{mean for A} - \text{mean for B} = 4.7 - 3.1 = 1.6$$

Calculate S:

$$S = \sqrt{\{\Sigma d^2 - [(\Sigma d)^2/n]\}/(n-1)}$$

where: Σd^2 = sum of the square of each difference

$= 0^2 + 1^2 + 3^2 + ... + 0^2 = 114$

$(\Sigma d)^2$ = sum of the differences, squared

$= 32^2 = 1024$

n = number of pairs = 20

therefore: $S = \sqrt{(114 - 1024/20)/19} = 1.82$

Find the t value from Statistical Chart 8 (Appendix) under the column headed 0.05.

The df is the number of pairs minus one:

$$\text{df} = 20 - 1 = 19$$

$$t \text{ value} = 2.093$$

The samples are significantly different if $\dfrac{\bar{d}}{S/\sqrt{n}} > t$

$$\frac{\bar{d}}{S/\sqrt{n}} = \frac{1.6}{1.82/\sqrt{20}} = 3.93$$

In this example 3.93 is greater than the t value, 2.093.

Using a paired comparison test as an alternative, we note there are 13 positive differences out of 15 nonzero differences. From Statistical Chart 3 the probability of this event under the null hypothesis is 0.007. Thus both test methods agree there is a difference in the two samples.

We conclude that cheese A was significantly more bitter than cheese B ($P \leq 0.05$).

Unstructured scaling

An alternative to the structured scale is an unstructured scale, also called line or visual analogue scales. In sensory analysis the unstructured scale most often used consists of a horizontal line 15 cm long with anchor points 1.5 cm from each end and often, but not necessarily, a mid point. Each anchor point is usually labeled with a word or expression. A separate line is used for each sensory attribute to be evaluated. Panelists record each evaluation by making a vertical line across the horizontal line at the point that best reflects their perception of the magnitude of that property (Fig. 5).

After the panelists have completed their judgments, the researcher measures the distance from the left end point of the line to each point marked by the panelist. The researcher then records the distances as intensity ratings between 0.0 and 15.0 for each product evaluated and analyzes these ratings.

Unstructured scales eliminate the problem of unequal intervals that is associated with structured scales. The terms to be used at the anchor points of the lines are agreed upon during panel training. These scales also represent a continuum and the comments under structured scaling apply.

A sample questionnaire and an example of descriptive analysis with an unstructured scale follow.

QUESTIONNAIRE FOR UNSTRUCTURED SCALING

PRODUCT: Wieners

NAME____________________________ DATE ____________________

Please evaluate the hardness and chewiness of these sample of wieners. Make a vertical line on the horizontal line to indicate your rating of the hardness and chewiness of each sample. Label each vertical line with the code number of the sample it represents.

Please taste the samples in the following order:

572 *681* *437* *249*

1. Hardness—the force required to compress the wiener sample between the molar teeth

—|——————————————————————|—

very soft very hard

2. Chewiness—the perceived work required to reduce the wiener sample to a consistency ready for swallowing

—|——————————————————————|—

slightly chewy very chewy

Comments:

Example This method was used when studying the texture of wieners. During preliminary testing the panelists determined that hardness and chewiness were the most important textural characteristics of wieners. The anchor words and definitions were agreed upon by the panel during preliminary sessions. Four brands of wieners were evaluated by eight panelists using the questionnaire.

Fig. 5 Tray prepared for an unstructured scaling test.

Numerical values were given to the ratings by measuring the distance of the panelists' marks from the left end of the line in units of 0.1 cm (Table 4). Analysis of variance was conducted on the numerical values for each characteristic.

Table 4 Hardness scores of four brands of wieners

	Treatments (Code)				
Panelist	A (249)	B (681)	C (437)	D (572)	Total
1	4.8	5.3	8.5	2.8	21.4
2	10.3	6.0	12.8	7.5	36.6
3	11.5	8.0	13.3	4.5	37.3
4	5.8	13.3	13.3	3.3	35.7
5	3.8	11.8	13.3	1.5	30.4
6	5.0	8.3	11.8	4.5	29.6
7	5.0	8.8	13.0	6.8	33.6
8	5.3	12.0	13.5	4.3	35.1
Total	51.5	73.5	99.5	35.2	259.7
Mean	6.44	9.19	12.44	4.40	

Analysis of variance

To complete an analysis of variance, certain calculations must be performed to determine the correction factor (CF), the sum of squares (SS), the degrees of freedom (df), the mean square (MS), and the variance ratio (or F value). A detailed explanation of these calculations using the results from Table 4 follows. This analysis can also be performed using a statistical package.

Correction factor The correction factor (CF) is calculated by squaring the total and dividing by the total number of judgments.

$$\begin{aligned} CF &= 259.7^2/(8 \times 4) \\ &= 67\,444.09/32 \\ &= 2107.63 \end{aligned}$$

Sum of squares (treatments) The sum of squares for treatments (SS(Tr)) is calculated by adding the squares of the total for each treatment, dividing by the number of judgments for each treatment, and then subtracting the correction factor.

$$\begin{aligned} SS(Tr) &= [(51.5^2 + 73.5^2 + 99.5^2 + 35.2^2)/8] - CF \\ &= [19\,193.79/8] - CF \\ &= 2399.22 - 2107.63 \\ &= 291.59 \end{aligned}$$

Sum of squares (panelists) The sum of squares for panelists (SS(P)) is calculated by adding the squares of the total for each panelist, dividing by the number of judgments made by each panelist, and subtracting the correction factor.

$$\begin{aligned} SS(P) &= [(21.4^2+36.6^2+37.3^2+35.7^2+30.4^2+29.6^2+33.6^2+35.1^2)/4] - CF \\ &= 8624.59/4 - CF \\ &= 2156.15 - 2107.63 \\ &= 48.52 \end{aligned}$$

Sum of squares (total) The sum of squares for the total (SS(T)) is calculated by adding the square of each judgment and subtracting the correction factor.

$$\begin{aligned} SS(T) &= (4.8^2 + 10.3^2 + 11.5^2 + 5.8^2 + 3.8^2 + 5.0^2 + 5.0^2 + \\ &\quad 5.3^2 + 5.3^2 + 6.0^2 + 8.0^2 + 13.3^2 + 11.8^2 + 8.3^2 + \\ &\quad 8.8^2 + 12.0^2 + 8.5^2 + 12.8^2 + 13.3^2 + 13.3^2 + 13.3^2 + \\ &\quad 11.8^2 + 13.0^2 + 13.5^2 + 2.8^2 + 7.5^2 + 4.5^2 + 3.3^2 + \\ &\quad 1.5^2 + 4.5^2 + 6.8^2 + 4.3^2) - CF \\ &= 2561.81 - 2107.63 \\ &= 454.18 \end{aligned}$$

Sum of squares (error) The sum of squares for error (SS(E)) is calculated by subtracting the SS values obtained from all specified sources of variation (treatments, panelists) from the SS for the total.

$$\begin{aligned} SS(E) &= 454.18 - 291.59 - 48.52 \\ &= 114.07 \end{aligned}$$

Degrees of freedom (treatments) The degrees of freedom for treatments (df(Tr)) are calculated by subtracting one from the number of treatments.

$$\begin{aligned} df(Tr) &= 4 - 1 \\ &= 3 \end{aligned}$$

Degrees of freedom (panelists) The degrees of freedom for panelists (df(P)) are calculated by subtracting one from the number of panelists.

$$\begin{aligned} df(P) &= 8 - 1 \\ &= 7 \end{aligned}$$

Degrees of freedom (total) The degrees of freedom for total (df(T)) are calculated by subtracting one from the total number of judgments.

$$\begin{aligned} df(T) &= 32 - 1 \\ &= 31 \end{aligned}$$

Degrees of freedom (error) The degrees of freedom for error (df(E)) are determined by subtracting the df for the other sources from the df for the total.

$$\begin{aligned} df(E) &= 31 - 7 - 3 \\ &= 21 \end{aligned}$$

Mean square The mean square (MS) for any variable is determined by dividing the sum of squares (SS) for that variable by its respective degrees of freedom (df).

$$\begin{aligned} MS(Tr) &= 291.59/3 \\ &= 97.20 \\ MS(P) &= 48.52/7 \\ &= 6.93 \\ MS(E) &= 114.07/21 \\ &= 5.43 \end{aligned}$$

Variance ratio (treatments) The variance ratio or F value for treatments (F(Tr)) is determined by dividing the MS for treatments by the MS for error.

$$\begin{aligned} F(Tr) &= 97.20/5.43 \\ &= 17.9 \end{aligned}$$

Variance ratio (panelists) The F value for panelists (F(P)) is determined by dividing the MS for panelists by the MS for error.

$$\begin{aligned} F(P) &= 6.93/5.43 \\ &= 1.3 \end{aligned}$$

The analysis of variance is summarized in Table 5.

Table 5 Analysis of variance of hardness of wieners

Source of variation	df	SS	MS	F
Treatments	3	291.59	97.20	17.9**
Panelists	7	48.52	6.93	1.3
Error	21	114.07	5.43	
Total	31	454.18		

**$P \leq 0.01$.

To determine if the difference among the treatments is significant, the calculated F value (17.9) is checked on Statistical Chart 9 (Appendix). With three degrees of freedom in the numerator and 21 degrees of freedom in the denominator, the variance ratio (F value) must exceed 3.07 to be significant at a probability of 0.05 (*) and must exceed 4.87 to be significant at $P \leq 0.01$ (**). The value 17.9 is therefore significant at the 0.01 probability level. We therefore conclude that there was a difference in hardness among the four brands of wieners.

Tukey's test

The treatments that are different from each other can be determined using Tukey's multiple comparison test (Snedecor and Cochran 1989). Multiple comparison tests (e.g., Scheffé, Tukey's, Newman-Keuls, Duncan's, and Fisher's LSD) are useful yardsticks for comparing means from qualitative treatments with no obvious order, such as those four brands of wieners. But multiple comparisons are not appropriate when

- treatments are levels of a quantitative variable, such as sugar added to coffee at 5, 10, or 15%;
- treatments are factorial combinations of factors, that is, each level of every factor occurs with each level of every other;
- comparisons of particular interest are noted at the planning stage.

For discussion of the alternatives to multiple comparison tests see Petersen (1977) or Little (1978).

The standard error of the treatment mean is calculated here using Table 5.

Standard error of the treatment mean The standard error of the treatment mean (SEM) is calculated by taking the square root of the MS for error divided by the number of judgments for each sample.

$$\begin{aligned} \text{SEM} &= \sqrt{5.43/8} \\ &= \sqrt{0.68} \\ &= 0.82 \end{aligned}$$

The significant studentized range value at $P = 0.05$ from Statistical Chart 7 (Appendix) for four treatments and 20 degrees of freedom is 3.96. For 21 degrees of freedom the interpolated value would be about 3.95.

Least significant difference The least significant difference (LSD) is calculated by multiplying the value obtained from the table by the SEM.

$$\begin{aligned} \text{LSD} &= 3.95 \times 0.82 \\ &= 3.24 \end{aligned}$$

Any two treatment means that differ by 3.24 or more are significantly different at $P \leq 0.05$.

Arrange the treatment according to magnitude.

C	B	A	D
12.44	9.19	6.44	4.40

Compare each mean with the others to see if the difference is 3.24 or more.

$$\begin{aligned} C - D &= 12.44 - 4.40 = 8.04 > 3.24 \\ C - A &= 12.44 - 6.44 = 6.00 > 3.24 \\ C - B &= 12.44 - 9.19 = 3.25 > 3.24 \end{aligned}$$

therefore C differs significantly from all the others.

$$\begin{aligned} B - D &= 9.19 - 4.40 = 4.79 > 3.24 \\ B - A &= 9.19 - 6.44 = 2.75 < 3.24 \end{aligned}$$

therefore B differs significantly from D but not from A.

$$A - D = 6.44 - 4.40 = 2.04 < 3.24$$

therefore A and D are not significantly different.

The results are shown using letters to indicate differences. Although additional decimal places are carried throughout to maintain accuracy, the results are presented to one decimal place. Any two values not followed by the same letter are significantly different at $P \leq 0.05$.

C	B	A	D
12.4a	9.2b	6.4bc	4.4c

Unstructured scaling with replication

The second example of unstructured scaling that follows is one of an analysis of variance with replication using a control. This example would be typical of data from a trained panel. Consumer evaluations are not replicated.

Example The juiciness of apples was evaluated. Each panelist received four apples, one for each of four different varieties (treatments) on each of three days (replicates). X was the control. The three new varieties were A, B, and C. The effects of treatments, panelists, replications, and their interactions were partitioned out (Table 6). For details of the method of analysis of variance see the previous example.

Table 6 Juiciness scores for apples

Panelist	Treatments X Replication 1	X 2	X 3	A Replication 1	A 2	A 3	B Replication 1	B 2	B 3	C Replication 1	C 2	C 3	Total
1	8.5	7.1	5.6	7.9	8.2	7.9	10.4	9.4	7.7	6.1	6.2	6.3	91.3
2	7.2	7.0	6.8	7.8	7.0	8.2	9.9	9.2	8.9	8.1	7.4	7.8	95.3
3	8.4	6.1	6.6	7.6	7.8	5.9	9.7	8.4	7.4	6.7	6.4	6.3	87.3
4	7.3	4.5	7.8	7.9	7.2	6.7	9.0	8.6	9.5	7.4	5.5	4:7	86.1
5	6.4	7.1	4.4	6.9	6.8	6.0	6.7	9.0	7.6	5.8	3.4	5.0	75.1
6	8.0	6.3	7.7	7.5	7.0	7.1	8.6	9.2	9.7	7.0	6.7	6.4	91.2
7	6.9	5.4	6.1	7.4	7.2	7.1	8.5	7.5	7.8	5.0	4.4	4.8	78.1
8	8.2	6.0	5.8	7.3	6.0	7.3	7.9	8.6	8.7	5.3	4.0	5.0	80.1
Treatment by replication total	60.9	49.5	50.8	60.3	57.2	56.2	70.7	69.9	67.3	51.4	44.0	46.3	684.5
Treatment total	X = 161.2			A = 173.7			B = 207.9			C = 141.7			
Replication total				1 = 243.3			2 = 220.6			3 = 220.6			

Correction factor:

$$\begin{aligned} CF &= \text{Total}^2/\text{number of responses} \\ &= (684.5^2)/(8 \times 4 \times 3) \\ &= 4880.63 \end{aligned}$$

Sum of squares (treatments):

$$\begin{aligned} SS(Tr) &= (\text{Sum of squares of the total for each treatment/number of judgments for each treatment}) - CF \\ &= [(161.2^2 + 173.7^2 + 207.9^2 + 141.7^2)/24] - CF \\ &= 4977.44 - 4880.63 \\ &= 96.81 \end{aligned}$$

Sum of squares (panelists):

$$\begin{aligned} SS(P) &= (\text{Sum of squares of the total for each panelist/number of judgments for each panelist}) - CF \\ &= [(91.3^2 + 95.3^2 + 87.3^2 + ... + 80.1^2)/12] - CF \\ &= 4910.45 - 4880.63 \\ &= 29.82 \end{aligned}$$

Sum of squares (replicates):

$$\text{SS(R)} = \text{(Sum of squares of the total for each replicate/number of judgments in each replicate)} - \text{CF}$$
$$= [(243.3^2 + 220.6^2 + 220.6^2)/32] - \text{CF}$$
$$= 4891.36 - 4880.63$$
$$= 10.73$$

Sum of squares (total):

$$\text{SS(T)} = \text{(Sum of squares of the total for each judgment)} - \text{CF}$$
$$= (8.5^2 + 7.2^2 + 8.4^2 + ... + 4.8^2 + 5.0^2) - 4880.63$$
$$= 5068.93 - 4880.63$$
$$= 188.30$$

Sum of squares (error):

$$\text{SS(E)} = \text{SS(T)} - \text{SS(Tr)} - \text{SS(P)} - \text{SS(R)}$$
$$= 188.30 - 96.81 - 29.82 - 10.73$$
$$= 50.94$$

The analysis of variance is summarized in Table 7.

Table 7 Analysis of variance of apple juiciness

Source of variation	df	SS	MS	*F*
Replicates	2	10.73	5.37	8.8**
Panelists	7	29.82	4.26	7.0**
Treatments	3	96.81	32.27	52.9**
Error	83	50.94	0.61	

**$P \leq 0.01$.

With three degrees of freedom in the numerator and 83 degrees of freedom in the denominator, the variance ratio (*F* value) must exceed an estimated 2.7 to be significant at $P \leq 0.05$ (*) and about 4.0 to be significant at $P \leq 0.01$ (**) (Statistical Chart 9 in Appendixes). The calculated *F* value is 52.9 for treatments. There is a significant difference at $P \leq 0.01$ (**).

There is also a significant panelist effect and a significant replicate effect at $P \leq 0.01$ (**). The significant panelist effect suggests the panelists were using different parts of the scale. It is important to determine if the panelists were scoring the samples consistently, that is, in the same order.

An analysis of variance with replication allows the sensory analyst to test for a panelist by treatment interaction. The absence of an interaction indicates the panelists are in agreement. To examine interactions in the data the following subtotals are required (Table 8).

Table 8 Treatment totals for each panelist

	Treatment			
Panelist	X	A	B	C
1	21.2	24.0	27.5	18.6
2	21.0	23.0	28.0	23.3
3	21.1	21.3	25.5	19.4
4	19.6	21.8	27.1	17.6
5	17.9	19.7	23.3	14.2
6	22.0	21.6	27.5	20.1
7	18.4	21.7	23.8	14.2
8	20.0	20.6	25.2	14.3

Sum of squares (treatment by panelist interaction):

$$\begin{aligned} SS(Tr \times P) &= \text{(Sum of squares of the treatment totals for each panelist/number of replications)} - SS(P) - SS(Tr) - CF \\ &= [(21.2^2 + 21.0^2 + 21.1^2 + \ldots + 14.2^2 + 14.3^2)/3] - 29.82 - 96.81 - 4880.63 \\ &= 5019.63 - 29.82 - 96.81 - 4880.63 \\ &= 12.37 \end{aligned}$$

Degrees of freedom (treatment by panelist interaction):

$$\begin{aligned} df(Tr \times P) &= df(Tr) \times df(P) \\ &= 3 \times 7 \\ &= 21 \end{aligned}$$

The interaction can then be added to the analysis of variance table. The sum of squares and degrees of freedom for error are as follows:

Sum of squares (error):

$$\begin{aligned} SS(E) &= SS(T) - SS(Tr) - SS(P) - SS(R) - SS(Tr \times P) \\ &= 188.30 - 96.81 - 29.82 - 10.73 - 12.37 \\ &= 38.57 \end{aligned}$$

Degrees of freedom (error):

$$\begin{aligned} df(E) &= df(T) - df(Tr) - df(P) - df(R) - df(Tr \times P) \\ &= 95 - 3 - 7 - 2 - 21 \\ &= 62 \end{aligned}$$

The replicate, treatment, panelist, and treatment by panelist interaction mean squares are now tested with the new error mean square (Table 9).

Table 9 Analysis of variance with replication of apple juiciness

Source of variation	df	SS	MS	F
Replicates	2	10.73	5.37	8.7**
Panelists	7	29.82	4.26	6.9**
Treatments	3	96.81	32.27	52.0**
Treatments × Panelists	21	12.37	0.59	1.0
Error	62	38.57	0.62	

$^{**}P \leq 0.01$.

The treatment by panelist interaction was not significant. We conclude that the panelists were in agreement. If this were not the case the panelists who do not agree are most easily spotted by plotting the treatment totals for each panelist against the treatment levels (Fig. 6). Inconsistent panelists will stand out as their scores will not be parallel to those of the other panelists.

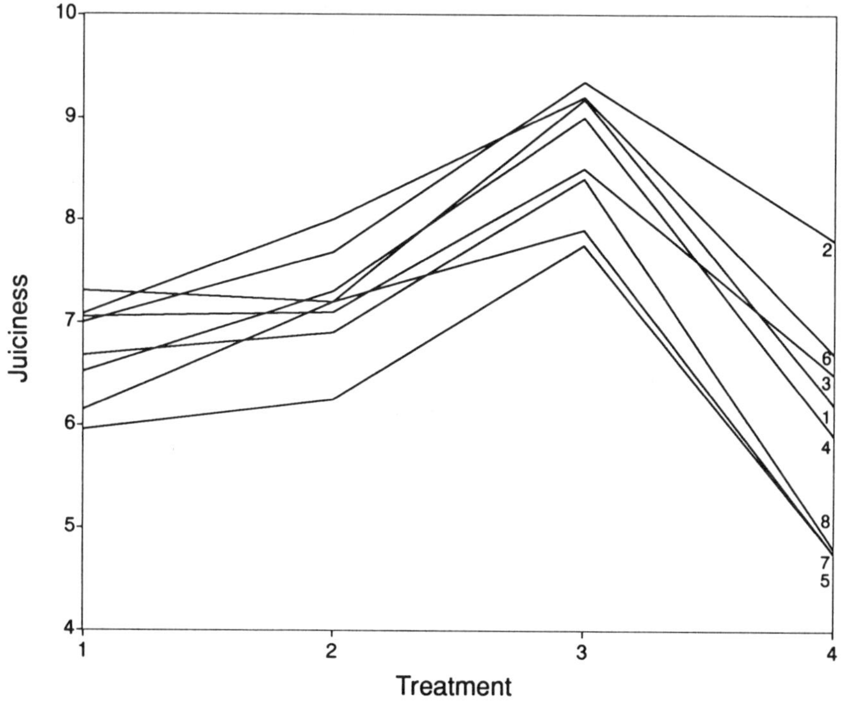

Fig. 6 Juiciness scores for panelists 1 to 8.

It is also possible to examine the replicate by treatment interaction and the replicate by panelist interaction in a similar fashion. A significant replicate by panelist interaction would indicate that some panelists were using different parts of the scale from replicate to replicate and this is not expected with a well-trained panel. A significant replicate by treatment interaction may indicate either that treatment effects vary over replicates or that the samples given to the different panelists are not independent. This could occur, for example, if all panelists were sampling from the same roasts. If this term is significant, the appropriate error term for testing treatment effects would be the replication by treatment interaction rather than SS(E), as SS(E) may underestimate the error. If not significant, it can be pooled with SS(E) as in the above analysis of variance. This technique of calculating interaction terms can also be used to calculate treatment A by treatment B interactions. For further discussion of interactions and choosing appropriate error terms see O'Mahony (1986). When any of these interactions are significant, give careful consideration to the interpretation of the data. To check treatment by replicate interaction, use the replication totals for each treatment as follows:

Sum of squares (treatment by replicate):

$$\begin{aligned} SS(Tr \times R) &= \text{(Sum of squares of the treatment totals for each replicate / number of panelists)} - SS(Tr) - SS(R) - CF \\ &= [(60.9^2 + 49.5^2 + 50.8^2 + \ldots + 46.3^2)/8] - 96.81 - 10.73 - 4880.63 \\ &= 4992.69 - 96.81 - 10.73 - 4880.63 \\ &= 4.52 \end{aligned}$$

Degrees of freedom (treatment by replicate):

$$\begin{aligned} df(Tr \times R) &= df(Tr) \times df(R) \\ &= 3 \times 2 \\ &= 6 \end{aligned}$$

Mean square (treatment by replicate):

$$\begin{aligned} MS(Tr \times R) &= SS(Tr \times R)/df(Tr \times R) \\ &= 4.52/6 \\ &= 0.75 \end{aligned}$$

To test this MS, the error MS must be calculated once again:

$$\begin{aligned} SS(E) &= \text{previous } SS(E) - SS(Tr \times R) \\ &= 38.57 - 4.52 \\ &= 34.05 \end{aligned}$$

$$\begin{aligned} df(E) &= \text{previous } df(E) - df(Tr \times R) \\ &= 62 - 6 \\ &= 56 \end{aligned}$$

$$\begin{aligned} MS(E) &= SS(E)/df(E) \\ &= 34.05/56 \\ &= 0.61 \end{aligned}$$

Calculate the F value for (Tr × R) :

$$F(\text{Tr} \times \text{R}) = \text{MS}(\text{Tr} \times \text{R})/\text{MS}(\text{E}) = 0.75/0.61 = 1.2$$

This F value with 6 degrees of freedom in the numerator and 56 degrees of freedom in the denominator is not greater than 2.3 (Statistical Chart 9 in Appendixes). Thus the treatment by replicate mean square is not significant and the previous analysis of variance is appropriate. If statistical software were used, you could include any interactions of interest in the analysis. Care must be taken to choose the correct error term.

Dunnett's test

The analysis of variance indicated a significant difference in the juiciness of the four treatments of apples. Variety X is the control treatment and the sensory analyst wishes to see if treatments A, B, and C differ from X. In the case where we wish to compare several treatments with a control, Dunnett's test is appropriate. This test can be either two tailed (e.g., are the other treatments different with respect to juiciness) or one tailed (e.g., are the other treatments juicier). For further discussion of one-tailed and two-tailed testing refer to O'Mahony (1986). In this case, we will use a one-tailed test. Calculations are very similar to Tukey's but a different table is used.

The standard error of the treatment means is calculated.

$$\text{SEM} = \sqrt{0.62/24} = \sqrt{0.026} = 0.16$$

The value from Statistical Chart 10 (Appendix), for four treatments (three excluding the control) and 60 degrees of freedom, is 2.10 ($P = 0.05$).

Calculate the least significant difference for $P = 0.05$ by multiplying the value obtained from the table by the square root of two and the SEM:

$$\text{LSD (0.05)} = 2.10 \times \sqrt{2} \times 0.16 = 0.48$$

Arrange sample means for treatments other than the control according to magnitude:

X	C	A	B
6.72	5.90	7.24	8.66

Compare each treatment mean with the control to see if it scored significantly more than the control. If the treatment scored 0.48 or more than the control it is significantly juicier at $P \leq 0.05$.

$$C - X = 5.90 - 6.72 = -0.82 < 0.48$$
$$A - X = 7.24 - 6.72 = 0.52 > 0.48$$
$$B - X = 8.66 - 6.72 = 1.94 > 0.48$$

The results can be shown as follows:

X	C	A	B
6.7	5.9	7.2*	8.7*

Thus we conclude that varieties A and B were juicier than the standard variety X at $P \leq 0.05$ (*).

Ratio scaling

Ratio scales are commonly used in physics. Scales of weight and distances are examples. A distance of 40 km is twice as long as a distance of 20 km. Ratios of measurements can be calculated. Ratio scaling is also used in sensory analysis. The ratio measurements are usually constructed by a procedure called magnitude estimation (Moskowitz 1988*a*). Panelists are given the samples arranged in random order that vary in one characteristic, such as hardness. They are instructed to assign any number to the first sample and to rate each sample in relation to the first. If the second sample seems twice as hard as the first, and if the first sample were assigned 50, the panelist would assign it the value 100; if it seems half as hard, it would be rated at 25.

Magnitude estimation is most appropriate for evaluating a quantity that does not include values near the threshold. Transform magnitude estimation data to logarithms before carrying out the appropriate analysis of variance (Butler et al. 1987). Because one cannot take the logarithm of zero, zero values pose a problem. However, if none of the products are near the threshold, zeros are unlikely. If zeros occur, a small constant may be added to all scores (Steel and Torrie 1980). A sample of the questionnaire and an example of magnitude estimation follow.

QUESTIONNAIRE FOR MAGNITUDE ESTIMATION

PRODUCT: Gelatin

NAME ______________________ DATE ______________

Please evaluate these samples of gel for hardness. Giving the first sample a value, assign relative values to each of the other samples to reflect their ratio to that of the first sample.

Samples	*Hardness*
649	______
872	______
259	______
138	______

Comments:

Example Magnitude estimation was used to find out if there was a difference in hardness of several samples of gel made with different amounts of gelatin (Table 10). Logarithms (natural) were taken and totals and means found (Table 11).

Table 10 Magnitude estimation scores of gel samples

Panelist	Treatment 1 (649)	2 (872)	3 (259)	4 (138)
1	50	100	150	200
2	150	400	600	700
3	100	200	300	400
4	75	75	90	100
5	6	8	10	11
6	100	150	200	300
7	150	100	200	300
8	50	70	85	100
9	30	60	100	120
10	50	60	100	125

Table 11 Magnitude estimates as logarithms

Panelist	Treatments 1 (649)	2 (872)	3 (259)	4 (138)	Total
1	3.912	4.605	5.011	5.298	18.826
2	5.011	5.991	6.397	6.551	23.950
3	4.605	5.298	5.704	5.991	21.598
4	4.317	4.317	4.500	4.605	17.739
5	1.792	2.079	2.303	2.398	8.572
6	4.605	5.011	5.298	5.704	20.618
7	5.011	4.605	5.298	5.704	20.618
8	3.912	4.248	4.443	4.605	17.208
9	3.401	4.094	4.605	4.787	16.887
10	3.912	4.094	4.605	4.828	17.439
Total	40.478	44.342	48.164	50.471	183.455
Mean	4.048	4.434	4.816	5.047	18.346

Analysis of variance

***NOTE:* Readers not familiar with the analysis of variance procedure should refer to the complete description given under "Unstructured scaling."**

Correction factor:

$$\begin{aligned} CF &= 183.455^2/40 \\ &= 841.393 \end{aligned}$$

Sum of squares (treatment):

$$\begin{aligned} SS(Tr) &= (40.478^2 + 44.342^2 + ... + 50.471^2)/10 - CF \\ &= 8471.77/10 - CF \\ &= 847.177 - 841.393 \\ &= 5.784 \end{aligned}$$

Sum of squares (panelists):

$$\begin{aligned} SS(P) &= (18.826^2 + 23.950^2 + ... + 17.439^2)/4 - CF \\ &= 3518.254/4 - CF \\ &= 879.564 - 841.393 \\ &= 38.171 \end{aligned}$$

Sum of squares (total):

$$\begin{aligned} SS(T) &= (3.912^2 + 5.011^2 + ... + 4.828^2) - CF \\ &= 886.713 - 841.393 \\ &= 45.320 \end{aligned}$$

The analysis of variance is summarized in Table 12.

Table 12 Analysis of variance of gel firmness

Source of variation	df	SS	MS	*F*
Treatment	3	5.784	1.928	37.8**
Panelists	9	38.171	4.241	83.2**
Error	27	1.365	0.051	

**$P \leq 0.01$.

The F-value for treatment is 37.8. According to Statistical Chart 9 (Appendix), if the F value exceeds 2.96 there is a significant difference at $P \leq 0.05$ (*); if it exceeds 4.40 there is a significant difference at $P \leq 0.01$ (**).

In this example the treatment levels were four ordered levels; 20, 25, 30, and 35 g of gelatin added to 800 mL of water or on the logarithmic scale 2.996, 3.219, 3.401, and 3.555. When the treatments are ordered levels it is often appropriate to look at a regression of the means on the treatment levels. In magnitude estimation, it has been postulated (Stevens 1956) that the treatments and levels are related as follows:

$$\log Y = \alpha + \beta \log X + \epsilon$$

where Y is the treatment mean
X is the treatment level
ϵ is the error
α and β are to be estimated.

To examine the linear regression of log Y on log X one proceeds on the logarithmic scale as follows.

Numerator correction factor The numerator correction factor (NCF) is the product of the sum of the treatment means and the sum of the treatment levels means all divided by the number of treatment levels.

$$\begin{aligned}\text{NCF} &= [(4.048 + 4.434 + 4.816 + 5.047) \times (2.996 + 3.219 + 3.401 + 3.555)]/4 \\ &= (18.345 \times 13.171)/4 \\ &= 241.6220/4 \\ &= 60.4055\end{aligned}$$

Numerator The numerator (NUM) is the sum of the cross product of treatment levels and treatment means minus the correction factor numerator, all multiplied by the number of panelists.

$$\begin{aligned}\text{NUM} &= [(2.996 \times 4.048) + (3.219 \times 4.434) + (3.401 \times 4.816) + (3.555 \times 5.047) - \text{NCF}] \times 10 \\ &= [60.7222 - 60.4055] \times 10 \\ &= 0.3167 \times 10 \\ &= 3.167\end{aligned}$$

Denominator correction factor The denominator correction factor (DCF) is the square of the sum of the treatment levels divided by the number of treatment levels.

$$\begin{aligned} \text{DCF} &= (2.996 + 3.219 + 3.401 + 3.555)^2/4 \\ &= 13.171^2/4 \\ &= 173.4752/4 \\ &= 43.3688 \end{aligned}$$

Denominator The denominator (DEN) is the sum of squares of the treatment levels minus the denominator correction factor all multiplied by the number of panelists.

$$\begin{aligned} \text{DEN} &= [(2.996^2 + 3.219^2 + 3.401^2 + 3.555^2) - \text{DCF}] \times 10 \\ &= [8.9760 + 10.3620 + 11.5668 + 12.6380 - 43.3688] \times 10 \\ &= 0.1740 \times 10 \\ &= 1.740 \end{aligned}$$

Slope The slope (β) is the numerator divided by the denominator, NUM/DEN.

$$\begin{aligned} \beta &= 3.167/1.740 \\ &= 1.82 \end{aligned}$$

Intercept The intercept (α) is the sum of the treatment means minus the slope times the sum of the treatment levels all divided by the number of levels.

$$\begin{aligned} \alpha &= (18.346 - 1.82 \times 13.171)/4 \\ &= (18.346 - 23.971)/4 \\ &= -5.625/4 \\ &= -1.41 \end{aligned}$$

Sum of squares (regression) The sum of squares for regression SS(R) is the numerator squared divided by the denominator.

$$\begin{aligned} \text{SS(R)} &= \text{NUM}^2/\text{DEN} \\ &= 3.167^2/1.74 \\ &= 5.764 \end{aligned}$$

Sum of squares (deviation from regression) The sum of squares for deviation from regression (SS(D)) is the sum of squares for treatment minus sum of squares for regression.

$$\begin{aligned} \text{SS(D)} &= \text{SS(Tr)} - \text{SS(R)} \\ &= 5.784 - 5.764 \\ &= 0.020 \end{aligned}$$

The analysis of variance can then be summarized as in Table 13.

Table 13 Analysis of variance of gel hardness

Source of variation	df	SS	MS	F
Treatments	3	5.784	1.928	37.8**
Regression log/log	(1)	5.764	5.764	113.0**
Deviation from regression	(2)	0.020	0.010	0.2
Panelists	9	38.171	4.241	83.2**
Error	27	1.365	0.051	

$^{**}P \leq 0.01$.

The F value of 113 for the regression of the log treatment effects on the log solution levels strongly indicates a linear effect on the log scale. The deviation from this line is nonsignificant indicating that the line fits very well (Fig. 7). The equation for the line is

$$\log Y = -1.41 + 1.82 \log X + \epsilon$$

This procedure is not limited to magnitude estimation but can be used whenever a linear relationship is postulated between the treatment scores and treatment levels. If the levels are equally spaced, the computations can be simplified by the use of orthogonal polynomials (Steel and Torrie 1980).

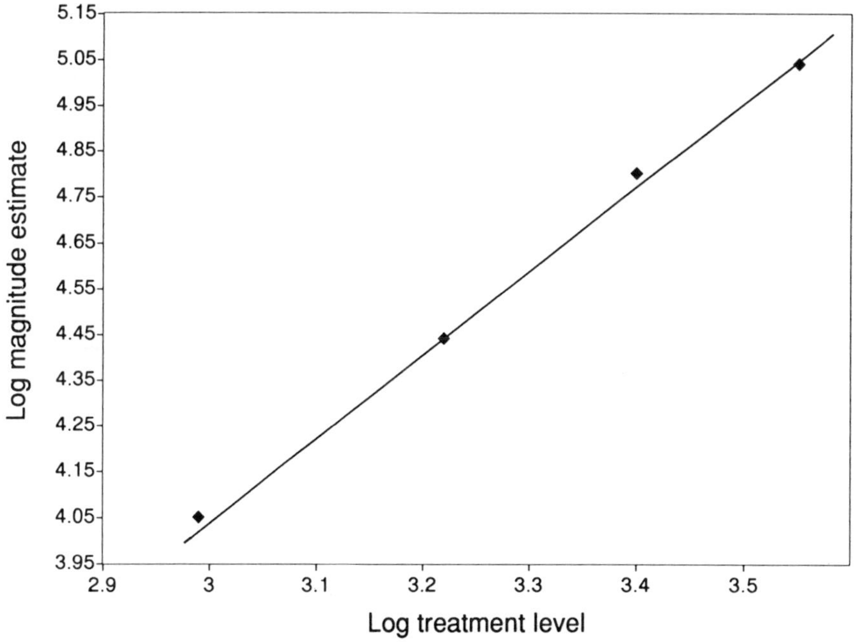

Fig. 7 Logarithm of the magnitude estimates plotted against logarithm of the treatment levels.

For more complicated relationships the reader is directed to Cochran and Cox (1957). Often a graph will be the best presentation of the results.

Descriptive analysis methods

The scaling tests described can be used to evaluate just one or several sensory characteristics of a product. Methods of descriptive analysis have been developed, which can be used to generate a more complete description or profile of the sensory quality of a product. Three such methods are the flavor profile, texture profile, and quantitative descriptive analysis.

Flavor profiling

The flavor profile method was introduced by Arthur D. Little Co., Cambridge, Massachusetts, in 1949 (Cairncross and Sjöström 1950; Caul 1957). The method provides a detailed, descriptive analysis of a product's flavor characteristics in both quantitative and qualitative terms. Trained panelists are used to analyze and discuss the flavor characteristics of a product in an open session approach to achieve a consensus. The final profile describes a product's aroma and flavor in terms of its detectable factors, their intensities, and their order of detection, any aftertaste, and an overall impression. The selection of panelists is based on taste and olfactory discrimination and descriptive ability (see "Selection and training of panelists"). During the long training process, panelists are trained in the fundamentals of the flavor profile method and in the physical and psychological aspects of tasting and smelling. They are presented with a wide selection of reference standards representing the product range, as well as samples to demonstrate ingredient and processing variables, to help develop and define the terminology that they will use.

During an actual flavor profiling session, four to six trained panelists sit around a table. The panelists first analyze the product or products individually, and then discuss their evaluations as a group. The products are analyzed one at a time for aroma, flavor, and mouth feelings, which are all called "character notes," using a degree of intensity scale that uses the following fairly broad demarcations:

0 = not present
)(= threshold
1 or + = slight
2 or + + = moderate
3 or + + + = strong

In some instances, for example, when they compare two very similar products, panelists can designate narrower ranges by using such symbols as 1/2, + (plus), or – (minus) (Caul 1957). The order of appearance of these character notes is indicated along with any aftertaste perceived.

An indication of the overall amplitude or impression of the aroma and flavor is given using the following scale:

)(	=	very low
1	=	low
2	=	medium
3	=	high

Because the final flavor profile of the product is a group consensus, no statistical analysis on the intensity values can be carried out. To circumvent this limitation, category scales or line scales can be used in place of the conventional 0,)(, 1, 2, 3 scale, and the panelists can make individual judgments rather than obtaining a group consensus.

An example of a scoresheet, which might be used in profiling the flavor of beer, follows.

QUESTIONNAIRE FOR FLAVOR PROFILE

PRODUCT: Beer

NAME ______________________ DATE ______________________

AROMA

Amplitude ________

	Intensity
Hoppy	________
Fruity	________
Sour	________
Yeasty	________
Malty	________

FLAVOR

Amplitude ________

	Intensity
Tingly (carbonation)	________
Sweet	________
Fruity	________
Bitter	________
Malty	________
Yeasty	________
Metallic	________
Astringent	________

AFTERTASTE ________

Comments:

During the group discussion, the panelists must reach a unanimous decision on the product evaluation. The panel leader then consolidates the panelists' conclusions into a concise description or flavor profile of the product.

Flavor profiles are often illustrated using a semicircular diagram (Fig. 8). The semicircle denotes the threshold concentrations, with the radiating lines corresponding to each individual character note (in order of appearance) and the length of the lines representing the intensity ratings (Cairncross and Sjöström 1950).

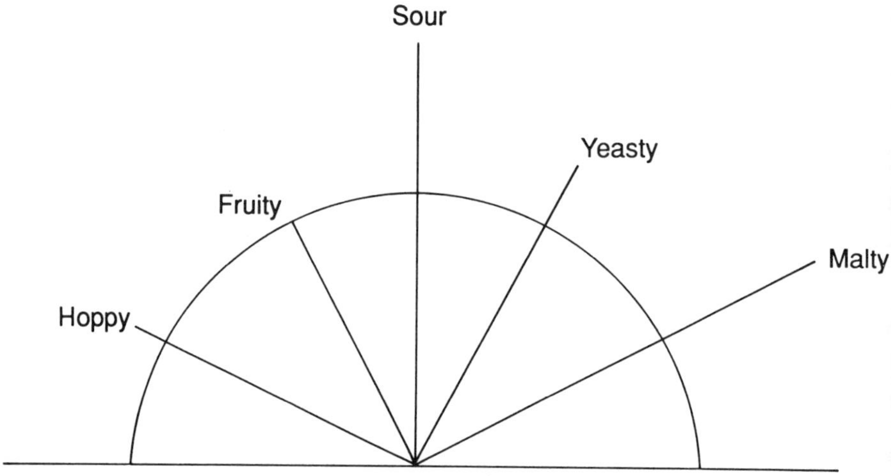

Fig. 8 Diagrammatic representation of the aroma portion of a beer sample flavor profile.

Texture profiling

The texture profile method was developed at the General Foods Research Center by Margaret Brandt and Alina Szczesniak (Brandt et al. 1963; Szczesniak 1963). This method classifies the textural parameters of a food not only into mechanical and geometrical qualities, but also into properties related to fat and moisture content. A quantitative and qualitative description is obtained with information on the intensity of each textural parameter present and the order in which they appear from the first bite through to complete mastication.

The selection of panelists is based on their textural discrimination and their descriptive ability. During training, panelists are introduced to the principles of texture as related to the structure of food. Through exposure to a wide range of food products, they are provided with a wide frame of reference for textural characteristics (Civille and Szczesniak 1973). Szczesniak and colleagues (1963) developed rating scales for different textural characteristics, which are useful during the panel training process.

The scales illustrate hardness, fracturability, chewiness, adhesiveness, and viscosity and include reference standards for each point on the scale. The reference foods are standardized with respect to brand name, handling procedure, sample size, and temperature. Substitutions of a reference food may be made, as long as the new food is a major brand of good consistency, requires minimum preparation to eliminate possible sources of variation, and does not change drastically with small temperature variations. Table 14 shows an example of the hardness scale.

Table 14 Standard hardness scale

Scale value	Product	Brand or type	Sample size	Temperature
1	Cream cheese	Kraft Philadelphia	1.5 cm cube	7–13 °C
2	Egg white	Hard-cooked (5 min)	1.5 cm of tip	20–25 °C
3	Frankfurters	Schneiders, large	1.5 cm slice	10–18 °C
4	Cheese	Kraft mild Cheddar	1.5 cm cube	10–18 °C
5	Olives	McLaren's, stuffed, queen-size	1 olive, pimento removed	10–18 °C
6	Peanuts	Cocktail-type	1 nut	20–25 °C
7	Carrots	Uncooked, fresh	1.5 cm slice	20–25 °C
8	Almonds	McNair, unblanched	1 nut	20–25 °C
9	Humbugs	McCormick	1	20–25 °C

Source: adapted from Szczesniak et al. (1963).

Each scale encompasses the entire range of intensity of the textural characteristic encountered in foods. The scale is first introduced in its entirety to familiarize the panelists with the specific texture parameter. Then the portion of the scale that corresponds to the extremes of the texture parameter of the test product or products is identified and expanded. The original texture profile method used an expanded 14-point version of the flavor profile scale. More recently, however, structured or unstructured scales, or magnitude estimation has been used. For example, the hardness of three new wieners is to be evaluated. Three points on the scale, 2 (egg white), 3 (frankfurters), and 4 (cheese), encompass the extremes in hardness of the three new test wieners. The three-point scale can be expanded by establishing reference points using wiener products between the two extreme points of egg white and cheese. This procedure is repeated for each texture attribute present in the product.

Originally, the texture profile method involved a group discussion and panel consensus as for the flavor profile method. However, now it is more common to have panelists evaluate the samples individually for each texture characteristic present using the developed scales to allow for statistical analysis of the data.

Quantitative descriptive analysis

A method of sensory analysis called quantitative descriptive analysis (QDA) was developed at the Stanford Research Institute (Stone et al. 1974) by which trained individuals identify and quantify the sensory properties of a product in order of occurrence. The basic features of the method are as follows:

- development of the sensory language as a group process
- panelist selection based on performance with test products
- as many as 12–16 repeat judgments from each panelist
- individual evaluations in booths
- unstructured scales
- analysis of variance to analyze individual and panel performance
- correlation coefficients to determine relationships among various scales
- statistical analysis to determine primary sensory variables
- multidimensional model developed and related to consumer responses.

An example of the attributes that might arise from QDA of orange jelly is shown in Table 15.

The type of visual display in Fig. 9 was suggested by Stone et al. (1974). The distance from the centre point to each attributes' point is the mean value of that attribute for each product. Standard errors could be included on the diagram.

Table 15 Results of analysis of variance[1] of orange jelly using quantitative descriptive analysis

Attribute	Brand A	Brand B	SEM[2]	Probability[3]
Orange color	10.2	7.9	0.62	0.011
Orange aroma	7.6	6.9	0.50	0.325
Firmness	9.6	6.6	0.64	0.001
Tartness	8.6	6.9	0.66	0.072
Orange flavor	7.6	6.9	0.72	0.494
Foreign flavor	4.3	4.8	0.48	0.464
Sweetness	7.1	9.6	0.42	< 0.001
Rate of breakdown	5.1	6.1	0.60	0.242

[1] Means based on 50 observations; 5 replicates of 10 panelists.
[2] Standard error of the mean based on 76 degrees of freedom.
[3] Probability that Brand A has the same intensity as Brand B.

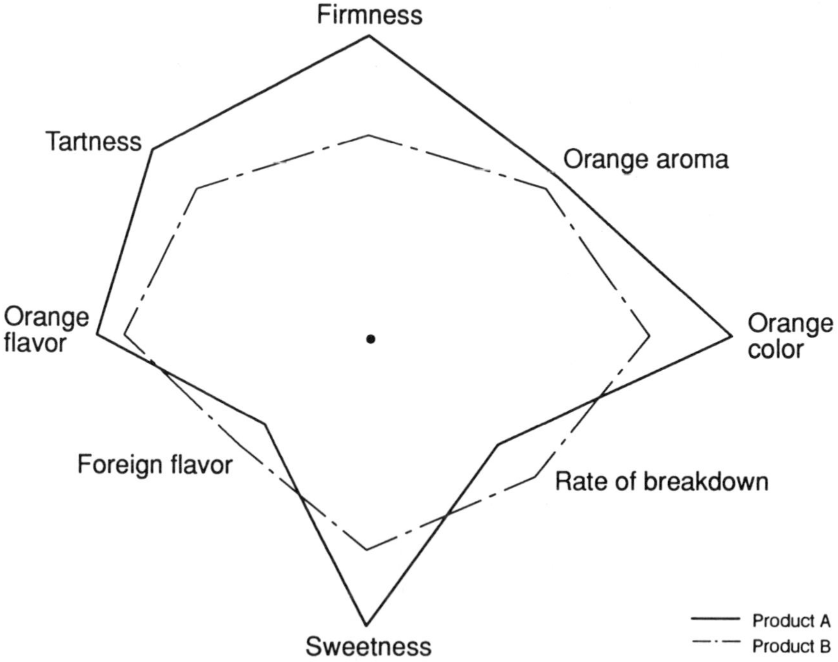

Fig. 9 Graphical representation of the orange jelly data.

Food Research Centre panel

Descriptive analysis at the Food Research Centre uses a form of attribute analysis for panel training and test product analysis. The following example outlines a panel procedure to analyze white sauces made with varying levels of a new, enriched flour. The objective is to examine the effect of flour substitution on the sensory quality of the sauce.

First, test several formulations to determine the best formulation for the control or standard white sauce and at the same time to standardize the cooking procedure. Once you have chosen the formulation, make sauces substituting varying levels of the new flour for the original flour; in this way determine if there is a maximum level for flour substitution based on physical performance. For example, at one substitution level the sauce might not thicken. Let us assume that after 25% substitution, the new flour does not allow the sauce to thicken. Therefore, the levels of new flour substitution to be examined are set at 0, 5, 15, and 25%. Once sample preparation is standardized, panel training can start.

During the first training session present the panelists with a control sauce (0% new flour substitution) and the 25% new flour sauce. Ask them to compare the sensory properties of the two samples and write down any differences and similarities on a blank piece of paper. The evaluations are done individually. After the session, the panel leader compares the responses and groups similar comments together, usually in the order of perception (i.e., visual, aroma, mouthfeel, taste/flavor, and aftertaste). For example, all the comments relating to visual properties are grouped together.

The next day, present the panelists with a list of all their comments from the previous day and any additional suggested terms from the panel leader. Then ask them to compare the control and 25% sauce again and, working with the list of provided comments as possible descriptors, to describe the differences and similarities between the two sauces. The panelists can add descriptors that are not on the list. Again, this evaluation is done individually. The panel leader then compares the comments and identifies any sensory descriptors commonly used amongst the panelists. These descriptors are the ones to be focused on during further training.

At the third training session, the panelists start working with the first grouping of sensory comments, in this case the visual descriptors. Present them with three sauces, a control, 5%, and 25%. Again using the list provided, the panelists make individual evaluations of the sauces. After each panelist has completed the evaluation, they discuss the observations as a group. Each descriptor is discussed separately with the hope of achieving agreement among the panelists as to which sample has the lowest and highest degree of that sensory characteristic. For example, if yellow color was a descriptor, the panel leader would try to establish consensus among the panelists as to which sauce was the most yellow, the least yellow and which is in between. The same procedure would take place for each sensory characteristic identified. Any characteristic that is noted by less than 50% of the panelists, during round-table discussions is dropped from the list if the panel leader agrees it is inappropriate. Otherwise, further training is needed to help the nonusers.

In preparation for the next day of training, the panel leader draws up a training ballot with a scale for each descriptor discussed the previous day. Examples are supplied to the panelists for any sensory characteristic for which a consensus was not reached. For example, Munsell color chips are useful color standards. The panelists discuss the standards and evaluate the test samples for each sensory characteristic on the ballot, followed by a group discussion of the samples. Once agreement is reached among panelists about the visual characteristics of the samples, the training moves on to the next grouping of sensory comments, in this case aroma.

Training for the aroma characteristics proceeds the same as described for the visual characteristics. Remember, during training the method of sample evaluation is also determined. For example, the aroma is to be evaluated by lifting the lid of the sauce container, taking three short sniffs, and then replacing the lid on the container. The lid is replaced to allow for headspace saturation in case the panelist wants to reevaluate a sample. Examples of problem descriptors are again supplied to help panelists to achieve consensus. One can use actual examples of the identified aroma, such as earth for an "earthy" aroma, or chemicals known to elicit certain perceived character notes, such as isopropyl quinoline for an "earthy" aroma. A useful guide to odorant chemicals is the *Atlas of odor character profiles* by A. Dravnieks (1985).

The next training sessions would focus on mouthfeel properties. By now the panelists are more familiar with the test samples, so you can include the fourth or 15% new flour substitution sample. Again, address and standardize the sample evaluation procedure. For example, standardize both the use of a different spoon for each sauce and the amount of sample taken into the mouth. A reference for texture terminology examples, which could be useful for panel training is Muñoz 1986.

Flavor would be the last grouping for which to train. When supplying samples to illustrate descriptors, keep in mind the preparation procedure of the test samples. For example, if "celery" is one of the descriptors panelists have identified, then the decision might be to supply both raw and cooked celery to the panelists because the sauce is cooked.

Once the panelists have been trained on all the appropriate sensory properties, the final scoresheet is put together by the panel leader. The order of appearance of each sensory characteristic on the scoresheet should be based on the order of perception and the logistics of evaluation. For example, although the panelists were trained on visual characteristics first, the aroma descriptors should appear first on the scoresheet owing to the importance of sample temperature and headspace saturation for aroma evaluation. An example of the final questionnaire follows.

QUESTIONNAIRE FOR WHITE SAUCE

NAME ______________________ DATE ______________________

Please evaluate the samples in the following order:

361 *478* *952* *660*

Aroma:

1) Earthy

slight ———————————————— intense

Color:

2) Yellow

slight ———————————————— intense

Mouthfeel:

3) Grainy

slight ———————————————— very

4) Consistency

thin ———————————————— thick

Flavor:

5) Buttery

slight ———————————————— intense

6) Salty

slight ———————————————— very

7) Celery

slight ———————————————— very

Comments:

This questionnaire is now tested. The panelists evaluate the test samples as if in a real test session. After the individual evaluations, the panelists discuss the samples and the evaluation procedure. For example: Are four samples too many? Are the descriptors in the correct order for evaluation? Are the appropriate words used for anchor points on the scales? At the same time, panelists are tested to see if they can reproduce their judgments. The reader is directed to American Meat Science Association (1978). Any necessary changes are made and retested before the real test sessions start.

After sufficient replication, the data are entered into the computer to be used for statistical analysis. Because the data are analyzed by computer, it is important to check for any errors in the data values that would more easily be detected were calculations done by hand. For each variable, maximums, minimums, means, and variances for each treatment, replicate, and panelist are examined. A careful examination of these helps us to detect any unusual values or problems in the data. For example, if one treatment has a very large or small variance we would not want to proceed to an analysis of variance which assumes variances are similar for all treatments. Plots of the data are also very useful at this stage. Once we are satisfied the data are in good order, the data are analyzed by the appropriate statistical method using a computer where possible. A report is prepared.

Affective tests

Affective tests are used to measure subjective attitudes towards a product based on its sensory properties. Test results give an indication of preference (select one over another), liking (degree of like/dislike), or acceptance (accept or reject) of a product (Pangborn 1980). The tests are generally used with a large number of untrained respondents to obtain an indication of the appeal of one product versus another.

Affective testing usually follows discriminative and descriptive testing, during which the number of product alternatives have been reduced to a limited subset. Stone and Sidel (1985) refer to the three primary types of affective tests as laboratory, central location, and home placement and suggest response numbers of 25–50, ≥100, and 50–100, respectively. The panelists are often selected to represent target markets.

Affective testing in the laboratory is used as part of the product screening effort to minimize testing of products that do not warrant further consideration. The laboratory panel can give an indication of product acceptability and provide direction for choosing products for the larger central location or home placement test. Three frequently used methods of affective testing are paired comparison, hedonic scaling, and ranking.

Paired comparison preference test

The paired comparison test used in preference testing is similar to that used in discriminative testing. The test requires the panelist to indicate which of two coded samples is preferred. Including a "no preference" or a "dislike both equally" option on the ballot is recommended only with a panel size of greater than 50 respondents (Gridgeman 1959; Stone and Sidel 1985). Permitting a tie with a small panel size reduces the statistical power of the test (i.e., reduces the probability of finding a difference between samples). Panelists are always concerned about making the right choice. They will often fall back on the "no preference" option, if it is included. Therefore, usually the panelists are asked to choose one sample, even if they perceive both samples as being the same, keeping the test as a forced choice test.

Two coded samples (A and B) are served simultaneously, with identical presentation style, i.e., same sample size, temperature, and container. There are two possible orders of presentation; A–B or B–A. Use each order an equal number of times for a small panel or select the order at random for a large panel. The order in which the panelist is to evaluate the samples is indicated on the ballot. Panelists usually evaluate only one pair of samples in a test with no replication. They are allowed to retaste the samples.

The researcher must decide if the test is a one- or two-tailed test. If the objective is to confirm a definite "improvement" or treatment effect on sample preference, then it is a one-tailed test. If the objective is to find which of two samples is preferred without any preconceived outcome, the test is a two-tailed test. The total number of panelists preferring each sample is calculated and tested for significance according to Statistical Chart 3 or 4 (Appendix). (See "Paired comparison test" for instructions on how to use the chart.) Although test results might indicate a preference for one sample over another, they give no data on the size of the difference in preference between the samples or on what the preference was based.

A sample questionnaire and an example of a paired comparison preference test follow.

QUESTIONNAIRE FOR PAIRED COMPARISON PREFERENCE TEST

PRODUCT: Cookies

NAME ______________________________ DATE ______________________

Taste the two cookies in the following order:

256 *697*

Which cookie do you prefer? You must make a choice.

____

Comments:

Example A paired comparison preference test was used to determine which of two chocolate-chip cookies was preferred (Fig. 10). Fifty panelists compared the two cookies. Twenty-five panelists evaluated a cookie from treatment A first, whereas the other 25 evaluated a cookie from treatment B first.

Fig. 10 Tray prepared for a paired comparison preference test.

Thirty-five of the 50 panelists preferred the cookies from treatment B. According to Statistical Chart 3 (Appendix), the probability is 0.007, which is less than the critical value of 0.05. The conclusion is that the cookie from treatment B was preferred by the panelists.

Hedonic scaling test

The most commonly used test for measuring the degree of liking of a sample is the hedonic scale. The term "hedonic" is defined as "having to do with pleasure." The scale includes a series of statements or points by which the panelist expresses a degree of liking or disliking for a sample. Scales of varying lengths can be used, but the most common is the 9-point hedonic scale, ranging from "like extremely" to "dislike extremely" with a central point of "neither like nor dislike" (Peryam and Girardot 1952).

The samples are coded and served in identical presentation style. The order of sample presentation is randomized for each panelist, and the order is indicated on the ballot. The samples can be served simultaneously or one at a time.

The responses are converted to numerical values ranging from 1 for "dislike extremely" to 9 for "like extremely." The data are analyzed either by *t*-test if only two samples are evaluated or by analysis of variance if three or more samples are evaluated. For a discussion of the appropriateness of the *t*-test and analysis of variance see "Structured scaling." An alternate analysis would be to rank the scores for panelists and conduct a Friedman's test (see "Ranking" under "Discriminitive tests").

A sample questionnaire and example of the 9-point hedonic scale follow.

QUESTIONNAIRE FOR HEDONIC SCALE

PRODUCT: Cottage cheese

NAME ______________________________ DATE ______________________

Please evaluate the four cottage cheese samples in the following order. Indicate how much you like or dislike each sample by checking the appropriate phrase.

216	*709*	*511*	*124*
__like extremely	__like extremely	__like extremely	__like extremely
__like very much	__like very much	__like very much	__like very much
__like moderately	__like moderately	__like moderately	__like moderately
__like slightly	__like slightly	__like slightly	__like slightly
__neither like nor dislike	__neither like nor dislike	__neither like nor dislike	__neither like nor dislike
__dislike slightly	__dislike slightly	__dislike slightly	__dislike slightly
__dislike moderately	__dislike moderately	__dislike moderately	__dislike moderately
__dislike very much	__dislike very much	__dislike very much	__dislike very much
__dislike extremely	__dislike extremely	__dislike extremely	__dislike extremely

Comments:

Example A 9-point hedonic scale was used to determine which brand of cottage cheese was most liked. Forty-one panelists evaluated the four samples. The data (Table 16) were submitted to analysis of variance (Table 17) to test for significance and Tukey's test was used to compare sample means (Table 18). (See "Unstructured scaling" for statistical analysis.)

Table 16 Hedonic scores for the four brands of cottage cheese

Panelist	Brand				Panelist	Brand			
	A	B	C	D		A	B	C	D
1	4	6	7	3	21	7	8	6	3
2	5	6	8	5	22	9	6	6	3
3	8	7	8	2	23	4	4	8	5
4	8	9	9	7	24	4	8	8	4
5	2	7	2	1	25	7	5	4	4
6	8	6	6	2	26	6	7	3	7
7	4	5	6	7	27	7	4	7	3
8	7	8	8	7	28	7	3	3	6
9	8	6	6	7	29	8	6	7	7
10	4	7	5	5	30	7	6	8	6
11	7	5	8	6	31	4	9	6	8
12	7	7	7	8	32	9	5	8	4
13	7	6	8	7	33	8	6	7	6
14	8	6	8	9	34	3	7	8	5
15	4	5	5	5	35	5	5	7	8
16	6	8	7	1	36	7	8	9	3
17	8	6	7	3	37	7	8	8	7
18	4	7	5	6	38	6	5	8	3
19	6	8	9	6	39	8	7	8	9
20	8	1	6	3	40	6	8	9	7
					41	5	5	6	8

Table 17 Analysis of variance of results of cottage cheese hedonic scales

Source of variation	df	SS	MS	*F*
Brands	3	50.39	16.80	5.69**
Panelists	40	183.99	4.60	1.56*
Error	120	354.11	2.95	
Total	163	588.49		

* $P \leq 0.05$; ** $P \leq 0.01$.

Table 18 Means and standard error of the mean (SEM) for the four brands of cottage cheese

Brand mean				
C	A	B	D	SEM
6.8 a	6.3 a	6.2 ab	5.3 b	0.27

The results indicate that the panelists liked brands C and A significantly more than brand D (Tukey's test). The mean scores of 6.8 to 6.3 for brand C and A, respectively, cover the "like moderately" (score = 7) and "like slightly" (score = 6) categories, whereas 5.3 for brand D corresponds to the "neither like nor dislike" (score = 5) category. Note that the difference between brands B and D is close to significance at $P = 0.05$.

Ranking test

The ranking test requires a panelist to evaluate three or more coded samples and to arrange them in ascending or descending order of preference or liking. Each sample must be assigned a rank; no ties are allowed. The panelist can be asked to rank for overall preference, or to zero in on a specific attribute, such as color or flavor preference.

Code and present the samples in identical style. Randomize the order of the samples for each panelist and indicate the order on the ballot. Serve the samples simultaneously to allow for any "among"-sample comparison necessary to assign ranks.

Total the ranks for each treatment and test for significance using a Friedman test for ranked data. Compare the differences between all possible pairs of ranks. (See "Ranking test (Friedman)" under "Discriminative tests" for further instructions.)

Although treatments will be ranked in ascending or descending order of preference or liking, the rank values do not indicate the amount or degree of difference between treatments. Also, because of the relative nature of the rank, values from one set of samples cannot be compared directly to another set of samples, unless both sets represent the same treatments. A sample questionnaire and example of the ranking test follow.

QUESTIONNAIRE FOR RANKING TEST

PRODUCT: Chocolate bars

NAME ______________________________ DATE ______________________

Please rank these chocolate bars in the order of acceptability. Rank the most acceptable chocolate bar as first and the least acceptable as fourth. Do not assign the same rank to two samples.

Evaluate the chocolate bars in the following order:

	551	*398*	*463*	*821*
	Rank		*Sample code*	
Most acceptable	First		________	
	Second		________	
	Third		________	
Least acceptable	Fourth		________	

Comments:

Example A ranking test was used to determine the order of acceptability for four chocolate bars with varying amounts of caramel. Forty panelists compared the samples. The rank sum for each chocolate bar is totaled.

The results are analyzed using the Friedman test for ranked data:

T = {12/[number of panelists × number of treatments × (number of treatments + 1)]} × (sum of the squares of the rank sum of each treatment) – 3 (number of panelists)(number of treatments + 1)

$= [12/(40)(4)(5)][41^2 + 84^2 + 127^2 + 151^2] - 3(40)(5)$

$= 115.01$

The calculated value of T is 115.01, which is greater than the value of χ^2 with 3 degrees of freedom for $\alpha = 0.05$, 7.81 (Statistical Chart 6 in Appendixes). Therefore we conclude that there is a significant difference in acceptability among the samples ($P \leq 0.05$). The least significant difference is determined using Statistical Chart 7 (Appendix) as described earlier (see "Ranking test (Friedman)" under "Discriminative tests").

$$\text{LSD rank} = 3.63\sqrt{[\text{No. panelists} \times \text{No. treatments} \times (\text{No. treatments} + 1)]/12}$$

$$= 29.6$$

Any two treatments where rank sums differ by more than 29.6 are significantly different (Table 19).

Table 19 Rank sum totals for the four chocolate bars

Chocolate bar	Rank sum[1]	Average rank
A	41a	1.0
B	84b	2.1
C	127c	3.2
D	151c	3.8

[1] Rank sums followed by the same letter are not significantly different ($P > 0.05$).

The results indicate that chocolate bar A was the most acceptable, chocolate bar B ranked second, whereas chocolate bars C and D were the least acceptable and did not differ between themselves.

Sensory analysis report

In any study or experiment, accurate and complete reporting is essential to the eventual usefulness of the results. Any report should contain enough detail

- to allow the reader to understand the study, to judge the appropriateness of the procedures, and to evaluate the reliability of the results
- to allow the study to be repeated
- to allow intra- and inter-laboratory comparisons to be made (Prell 1976).

In preparing a report follow these guidelines: title, abstract or summary, introduction, experimental method, results and discussion, conclusions, and references. For more information and actual examples, refer to Prell (1976), Larmond (1981), and Meilgaard et al. (1987*b*).

Title

The first information necessary in a report is the title of the project or experiment, the names of the persons who are responsible for reporting the work, their affiliations, and when the work was done.

Abstract or summary

If the report is for publication in a journal, include a short abstract or summary, generally of 100–200 words in length (the length is specified by the journal). In the abstract, state the objective, provide a concise description of the experiment or experiments, and report the major observations, the significance of the results, and the conclusions. Even if the report is not targeted for publication, a brief summary can be useful to the reader, particularly nontechnical readers, such as managers.

Introduction

In the introduction, clearly state the aim of the project as well as the objective of each test within that project. Define the purpose of the investigation or the problem to be solved, e.g., new product development, product matching, product improvement, storage stability, and so on. Review or cite any pertinent previous work.

Experimental method

Under experimental method, describe the sensory procedures and statistical analyses used. Give sufficient detail about the method and equipment to allow the work to be repeated. Always cite accepted methods by appropriate and complete references. The use of subheadings here can help to provide clarity, which makes information more useful. Consider the following subheadings:

Experimental design Include here the statistical design used (e.g., randomized complete block, incomplete block, or split plot); the measurements made (i.e., sensory, chemical, and physical); factors and levels of factors; and number of replications. State any limitations to the design, such as only certain lots being available for sampling.

Sensory method Identify the sensory method or methods used and give appropriate references, such as the International Organization for Standardization (ISO), American Society for Testing and Materials (ASTM), or papers from refereed journals.

Sensory panel State the source of the panel (in-house or recruited from outside the organization) and number of panelists. If the panelists were trained, give details on the method of selection and training. Information on the composition of the panel, such as age and sex, is usually important when affective tests are used.

Environmental conditions Describe the test location (i.e., laboratory, shopping mall, or home) and lighting. Include other information, such as room temperature or existence of distractions (e.g., odors or noise).

Sample preparation and presentation Provide details on the equipment for, and method or methods of, sample preparation (e.g., electric oven, time, and temperature). Specify the use of sample codes (i.e., three-digit random numbers), order of presentation, sample size, carrier, temperature, container, utensils, time of day, special instructions to panelists, time intervals, rinsing agent(s), whether samples were swallowed or expectorated, and any other conditions that were controlled or would influence the data collected.

Statistical techniques Describe the manner in which numbers or scores were derived from the test responses to enable data analysis. Discuss the type of statistical analysis used.

Results and discussion

Present results clearly and concisely, summarizing the relevant collected data, but giving enough data to justify conclusions. When reporting tests of significance, indicate the probability level, degrees of freedom, calculated value of the test (F, χ^2, t, etc.), and direction of the effect. Besides words, use either tables or figures to present results, but avoid presenting the same information twice. Tables can also be used to report analysis of variance results, or treatment means and their standard errors.

Data are often more easily understood and discussed if they can be visualized through the use of charts and graphs. Fig. 11 is an example of a frequency distribution presented as a bar graph or histogram. If only the mean score of 6, corresponding to "like slightly" on the hedonic scale, is

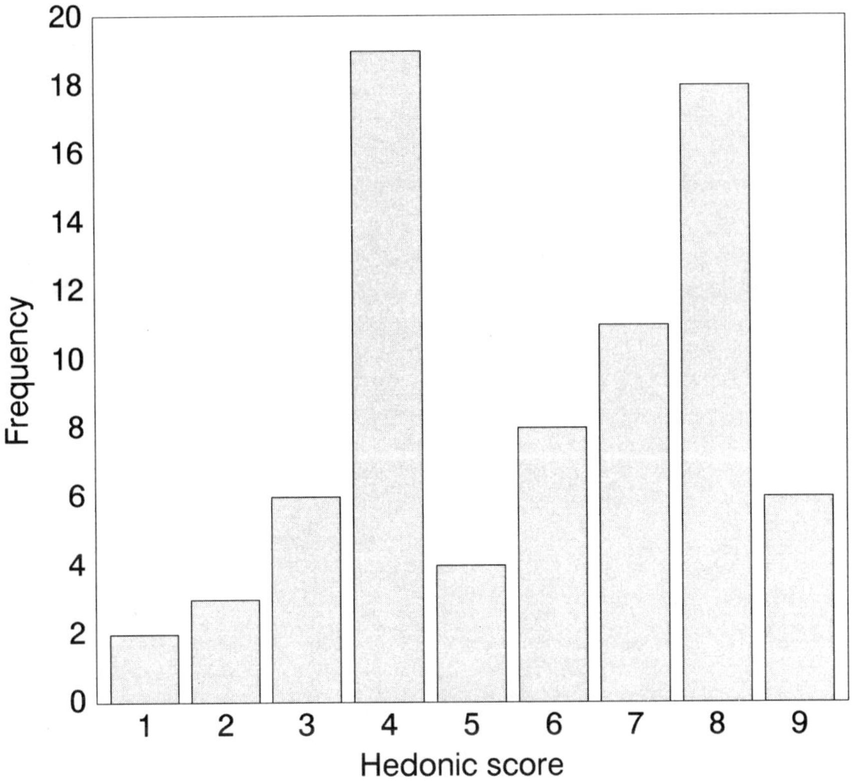

Fig. 11 Example of a histogram.

reported, one does not see the whole picture. The histogram shows that there are actually two different groups of consumers—one group who most often rated the product as "dislike slightly," a score of 4, and a second group who most often rated the product as "like very much," a score of 8. In this case, visualization of the results gives more information than just reporting the mean.

Fig. 12 is an example of results presented in graphical form. A 10-member trained panel evaluated the perceived intensity of sourness in lemonade with varying levels of sucrose added. The mean sourness scores are plotted against the sucrose levels. The standard error bars are included for an indication of variability. The graph makes it easy to see the sharp decrease in sourness perception from 0 to 4% sucrose added, with the much more gradual decrease from 4 to 8% sucrose added.

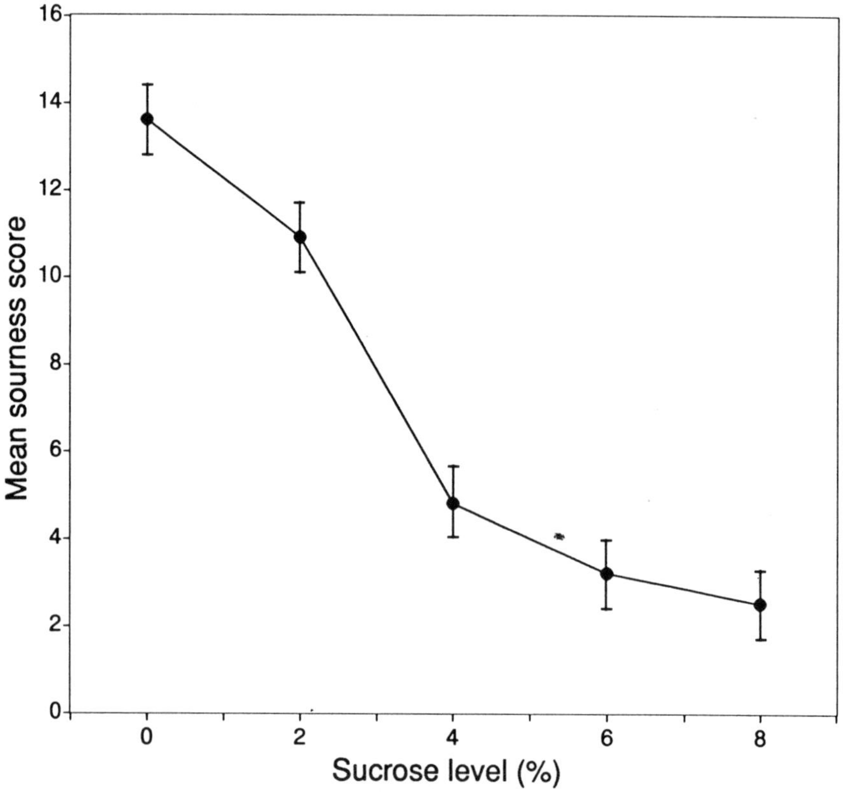

Fig. 12 Example of graphical form for presentation of results.

These are just a few examples of ways to present results. It might be necessary to try a few different ways to decide which is the best for the report. Remember to keep the table, chart, or graph simple and uncluttered. No matter in what form data are presented, properly tabulated or graphed data can make it easier for the researcher to understand the results of an experiment and can also aid in the communication of the results to others.

The results should be "interpreted, compared, and contrasted" (Prell 1976). The discussion should identify both the theoretical and practical significance of the results and should relate the new findings to any previous results, if possible. It is best to discuss the results in the same order as the study was carried out.

Conclusions

Include a final paragraph with the conclusions drawn from the study in the report. Keep the conclusions brief and clearly stated. Make any recommendations for further study at this point.

References

If references are used and cited, give a complete list to enable the reader to locate a desired citation. Refer to specific journal guidelines for the information required and an example of presentation format.

References

American Meat Science Association. 1978. Guidelines for cookery and sensory evaluation of meat. AMSA, Chicago, Ill. 24 pp.

American Society for Testing and Materials. 1968. Manual on sensory testing methods. ASTM STP 434. American Society for Testing and Materials, Philadelphia, Pa. 77 pp.

American Society for Testing and Materials. 1981. Guidelines for the selection and training of sensory panel members. ASTM STP 758. American Society for Testing and Materials, Philadelphia, Pa. 35 pp.

American Society for Testing and Materials. 1986. Physical requirement guidelines for sensory evaluation laboratories. ASTM STP 913. Eggert, J.; Zook, K., eds. American Society for Testing and Materials, Philadelphia, Pa. 54 pp.

Amerine, M.A.; Pangborn, R.M.; Roessler, E.B. 1965. Principles of sensory evaluation of food. Academic Press, New York, N.Y. 602 pp.

Brandt, M.A.; Skinner, E.; Coleman, J. 1963. Texture profile method. J. Food Sci. 28:404–410.

Butler, G.; Poste, L.M.; Wolynetz, M.S.; Agar V.E.; Larmond, E. 1987. Alternative analyses of magnitude estimation data. J. Sensory Stud. 2:243–257.

Cairncross, W.E.; Sjöström, L.B. 1950. Flavor profile—A new approach to flavor problems. Food Technol. 4:308–311.

Caul, J.F. 1957. The profile method of flavor analysis. Adv. Food Res. 7:1–40.

Civille, G.V.; Szczesniak, A.S. 1973. Guidelines to training a texture profile panel. J. Texture Stud. 4:204–233.

Cochran, W.G.; Cox, G.M. 1957. Experimental design. John Wiley and Sons, Inc., New York, N.Y. 611 pp.

Dravnieks, A. 1985. Atlas of odor character profiles. ASTM DS 61. American Society for Testing and Materials. Philadelphia, Pa. 354 pp.

Dunnett, C.W. 1955. A multiple comparison procedure for comparing several treatments with a control. J. Am. Stat. Assoc. 50:1096–1121.

Dunnett, C.W. 1964. New tables for multiple comparisons with a control. Biometrics 20:482–491.

Ellis, B.H. 1961. A guide book for sensory testing. Continental Can. Co., Chicago, Ill. 55 pp.

Fisher, R.A.; Yates, F. 1974. Statistical tables for biological, agricultural and medical research. 6th edition. Oliver and Boyd Ltd., Edinburgh. 112 pp.

Gacula, M.C. 1978. Analysis of incomplete block designs with reference samples in every block. J. Food Sci. 43:1461–1466.

Gacula, M.C., Jr.; Singh, J. 1984. Statistical methods in food and consumer research. Academic Press, Orlando, Fla. 505 pp.

Gridgeman, N.T. 1959. Pair comparison, with and without ties. Biometrics 15:382–388.

Hollander, M.; Wolfe, D.A. 1973. Non parametric statistical methods. John Wiley and Sons Inc., New York, N.Y. 503 pp.

Institute of Food Technologists. 1975. Minutes of Sensory Evaluation Div. business meeting at 35th Ann. Meet., Inst. of Food Technologists, Chicago, June 10.

International Organization for Standardization. 1985. Sensory analysis—General guidance for the design of testing facilities. DP 8589.

Kroll, B.J.; Pilgrim, F.J. 1961. Sensory evaluation of accessory foods with and without carriers. J. Food Sci. 26:122–124.

Larmond, E. 1981. Better reports of sensory evaluation. Tech. Q. Master Brew. Assoc. Am. 18(1):7–10.

Little, T.M. 1978. If Galileo published in Hort. Science. Hortic. Sci. 13:504–506.

Marshall, S.G.; Vaisey, M. 1972. Sweetness perception in relation to some textural characteristics of hydrocolloid gels. J. Texture Stud. 3:173–185.

Meilgaard, M.; Civille, G.V.; Carr, B.T. 1987*a*. Sensory evaluation techniques. Vol. I, CRC Press, Inc., Boca Raton, Fla. 159 pp.

Meilgaard, M., Civille, G.V.; Carr, B.T. 1987*b*. Sensory evaluation techniques. Vol. II, CRC Press, Inc., Boca Raton, Fla. 159 pp.

Moskowitz, H. 1988*a*. Applied sensory analysis of foods. Vol. I, CRC Press, Inc., Boca Raton, Fla. 259 pp.

Moskowitz, H. 1988*b*. Applied sensory analysis of foods. Vol. II, CRC Press, Inc., Boca Raton, Fla. 180 pp.

Muñoz, A. 1986. Development and application of texture reference scales. J. Sensory Stud. 1:55–83.

Newell, G.J.; MacFarlane, J.D. 1987. Expanded tables for multiple comparison procedures in the analysis of ranked data. J. Food Sci. 52(6):1721–1725.

O'Mahony, M. 1986. Sensory evaluation of food: Statistical methods and Procedures. Tannenbaum, S.R.; Walstra, P. eds. Marcel Dekker, New York, N.Y. 487 pp.

Pangborn, R.M. 1980. Sensory science today. Cereal Foods World 25(10): 637–640.

Pangborn, R.M.; Szczesniak, A.S. 1974. Effect of hydrocolloids and viscosity on flavour and odour intensities of aromatic flavour compounds. J. Texture Stud. 4:467–482.

Peryam, D.R.; Girardot, N.F. 1952. Advanced taste test method. Food Eng. (N.Y.) 24:58–61.

Petersen, R. 1977. Use and misuse of multiple comparison procedures. Agron. J. 69:205–208.

Prell, P.A. 1976. Preparation of reports and manuscripts which include sensory evaluation data. Food Technol. November 40–46.

Roessler, E.B.; Pangborn, J.I.; Sidel, J.L.; Stone, H. 1978. Expanded statistical tables for estimating significance in paired-preference, paired difference, duo-trio and triangle tests. J. Food Sci. 43:940–943.

Snedecor, G.W.; Cochran, W.G. 1989. Statistical methods. 8th edition. Iowa State University Press, Ames, Iowa. 507 pp.

Steel, R.G.D.; Torrie, J.H. 1980. Principles and procedures of statistics. A biometrical approach. 2nd edition. McGraw-Hill Book Co. New York, N.Y. 633 pp.

Stevens, S.S. 1956. The direct estimation of sensory magnitude—loudness. Am. J. Psychol. 69:1–25.

Stone, H.; Sidel, J.L. 1985. Sensory evaluation practices. Schweigart, B.S.; Stewart, G.F. eds. Academic Press, Inc., Orlando, Fla. 311 pp.

Stone, H.H.; Sidel, J.; Oliver, S.; Woolsey, A.; Singleton, R.C. 1974. Sensory evaluation by quantitative descriptive analysis. Food Technol. 28(11):24–31.

Szczesniak, A.S. 1963. Classification of textural characteristics. J. Food Sci. 28:385–389.

Szczesniak, A.S.; Brandt, M.A.; Friedman, H.H. 1963. Development of standard rating scales for mechanical parameters of texture and correlation between the objective and the sensory methods of texture evaluation. J. Food Sci. 28:397–403.

APPENDIXES

Statistical Chart 1 Table of random numbers, permutations of nine

98119	47634	62128	74824	26316	69967	99242
42293	62781	39637	56945	93661	35153	26837
71926	19563	58873	41611	12194	24228	17798
17455	58857	11764	19452	57975	47815	52523
66834	25245	27285	25299	71782	88679	34114
29662	83196	93516	32777	64843	92581	73375
35341	74918	44949	93188	85428	71742	68486
84787	96479	76491	68563	38259	16396	81659
53578	31322	85352	87336	49537	53434	45961
24814	99952	56378	32381	21148	97297	72848
52498	87383	22131	69919	15466	36866	98794
71675	78536	73495	27778	77622	55159	51263
98581	33164	49769	86257	88255	72928	85426
37137	45775	97913	51435	93913	14312	16975
63226	16691	38586	93122	54874	29581	44112
86349	64849	61622	15644	62331	63645	39681
19752	22217	85847	48893	46597	48774	63557
45963	51428	14254	74566	39789	81433	27339
35951	41886	65455	39863	86891	26531	12772
62737	28622	97744	94399	65615	71385	36396
93289	53491	21397	16212	98532	12463	83527
57693	75743	49661	77747	54358	44124	59213
74862	87515	54246	43585	19779	87859	64959
19174	39259	33113	82151	47924	59642	75834
86518	66168	12838	25974	31147	93998	41141
48426	14937	88522	68628	73483	38217	97468
21345	92374	76989	51436	22266	65776	28685
46622	22565	57564	62716	48346	22575	76356
32234	77979	61291	19143	19222	98313	54127
25553	66127	89656	47931	84577	46781	43718
14871	33813	48879	33229	37769	79698	38599
91485	98384	23448	75582	95118	67862	82865
57997	15656	36185	56864	21884	53946	61484
78366	84292	74722	28657	76693	84137	99632
63148	49438	15937	81498	52455	31229	17273
89719	51741	92313	94375	63931	15454	25941
85392	17996	58885	38247	84138	71165	44722
72575	99477	91117	93856	77347	82872	29147
51457	72341	72394	47919	62519	34731	82898
96724	46815	23931	75785	95794	15923	57213
48683	28624	46578	52168	11983	99488	61586
64968	51183	64763	19332	33622	27299	73355
27846	64569	85256	81471	49461	58617	95634
19211	35232	19449	26624	58256	66356	18461
33139	83758	37622	64593	26875	43544	36979

(continued)

Statistical Chart 1 (*concluded*)

42659	14978	74643	21224	33681	47164	99323
68131	96442	21839	59659	22718	79895	24254
36478	75184	92278	74478	19924	63749	61566
14824	37556	35982	63737	45539	56252	46132
79966	61713	57561	85393	54495	38978	73771
95313	43327	16415	42881	97242	84586	58488
21585	82635	43154	16545	86363	91337	82695
53797	28899	69727	38112	61157	12611	37847
87242	59261	88396	97966	78876	25423	15919
16383	72679	78165	11448	12781	89769	75817
22898	25526	34851	48721	65122	95142	39438
87751	97953	43779	55296	27956	53894	62683
35614	36891	15414	87175	88867	72978	84774
93275	41737	91937	24934	79478	36256	17991
44946	18212	22682	79363	46615	24425	26256
78167	54368	69326	36587	34349	68617	91365
61439	89445	56293	92612	91594	41581	58129
59522	63184	87548	63859	53233	17333	43542
49661	11831	37549	97499	94883	32513	95688
53196	68259	65492	28563	21942	86426	81796
86857	75113	73927	69736	86511	95998	73811
37442	22945	91338	12117	39629	48254	12377
98339	49672	86783	81928	17356	53331	29969
24714	33386	29114	36371	42134	69875	54255
15928	96568	14661	55252	75467	11189	47432
71583	54794	58875	74885	53795	27767	38544
62275	87427	42256	43644	68278	74642	66123
31727	54363	98644	86696	58126	54111	12173
22588	96555	31488	39317	73757	67449	37334
15869	22124	49991	13468	84674	28392	89592
63276	85881	75722	45251	12565	72976	44247
98414	73738	64539	57729	36299	46527	76481
57633	41279	52277	94144	21331	19263	23856
46155	17446	13115	68983	67448	33855	98668
84991	39692	86853	21575	45912	85738	51719
79342	68917	27366	72832	99883	91684	65925
92495	48448	19485	27965	98734	38213	35326
11813	86599	27677	68698	22229	14862	28984
39557	24933	81923	76577	67867	25957	14118
86672	69156	96531	11751	83458	93428	51672
75339	73687	68254	34146	59592	62575	69737
24921	97224	42748	83432	46971	77696	72261
43784	51812	73399	99219	31113	89734	43549
68166	35771	55116	52383	15686	46389	86495
57248	12365	34862	45824	74345	51141	97853

Statistical Chart 2 Triangle test, probability of x or more correct judgments in n trials (one-tailed, $p = 1/3$)[a]

$n \backslash x$	0	1	2	3	4	5	6	7	8	9	10	11	12	13	14	15	16	17	18	19	20	21	22	23	24	25	26	27	28
5		868	539	210	045	004																							
6		912	649	320	100	018	001																						
7		941	737	429	173	045	007																						
8		961	805	532	259	088	020	003																					
9		974	857	623	350	145	042	008	001																				
10		983	896	701	441	213	077	020	003																				
11		988	925	766	527	289	122	039	009	001																			
12		992	946	819	607	368	178	066	019	004	001																		
13		995	961	861	678	448	241	104	035	009	002																		
14		997	973	895	739	524	310	149	058	017	004	001																	
15		998	981	921	791	596	382	203	088	031	008	002																	
16		998	986	941	834	661	453	263	126	050	016	004	001																
17		999	990	956	870	719	522	326	172	075	027	008	002																
18		999	993	967	898	769	588	391	223	108	043	014	004	001															
19			995	976	921	812	648	457	279	146	065	024	007	002															
20			997	982	940	848	703	521	339	191	092	038	013	004	001														
21			998	987	954	879	751	581	399	240	125	056	021	007	002														
22			998	991	965	904	794	638	460	293	163	079	033	012	003	001													
23			999	993	974	924	831	690	519	349	206	107	048	019	006	002													
24			999	995	980	941	862	737	576	406	254	140	068	028	010	003	001												
25			999	996	985	954	888	778	630	462	304	178	092	042	016	006	002												
26				997	989	964	910	815	679	518	357	220	121	058	025	009	003	001											
27				998	992	972	928	847	725	572	411	266	154	079	036	014	005	002											
28				999	994	979	943	874	765	623	464	314	191	104	050	022	008	003	001										
29				999	996	984	955	897	801	670	517	364	232	133	068	031	013	005	001										
30				999	997	988	965	916	833	714	568	415	276	166	090	043	019	007	002	001									
31					998	991	972	932	861	754	617	466	322	203	115	059	027	011	004	001									
32					998	993	978	946	885	789	662	516	370	243	144	078	038	016	006	002	001								
33					999	995	983	957	905	821	705	565	419	285	177	100	051	023	010	004	001								
34					999	996	987	965	922	849	744	612	468	330	213	126	067	033	014	006	002	001							
35					999	997	990	973	937	873	779	656	516	376	252	155	087	044	020	009	003	001							
36						998	992	978	949	895	810	697	562	422	293	187	109	058	028	012	005	002	001						
37						998	994	983	959	913	838	735	607	469	336	223	135	075	038	018	007	003	001						
38						999	996	987	967	928	863	769	650	515	381	261	164	095	051	025	011	004	002	001					
39						999	997	990	973	941	885	800	689	560	425	301	196	118	066	033	016	007	003	001					
40						999	997	992	979	952	903	829	726	603	470	342	231	144	083	044	021	010	004	001					
41							998	994	983	961	920	854	761	644	515	385	268	173	104	057	029	014	006	002	001				
42							999	995	987	968	933	876	791	683	558	428	307	205	127	073	038	019	008	003	001				
43							999	996	990	974	945	895	820	719	600	471	347	239	153	091	050	025	012	005	002	001			
44							999	997	992	980	955	912	845	753	639	514	389	275	182	111	063	033	016	007	003	001			
45							999	998	994	984	963	926	867	783	677	556	430	313	213	135	079	043	022	010	004	002	001		
46								998	995	987	970	938	887	811	713	596	472	352	246	161	098	055	029	014	006	003	001		
47								999	996	990	976	949	904	836	745	635	514	392	282	189	119	070	038	019	009	004	002	001	
48								999	997	992	980	958	919	859	776	672	554	433	318	220	142	086	048	025	012	006	002	001	
49								999	998	994	984	965	932	879	803	706	593	473	356	253	168	105	061	033	017	008	003	001	
50								999	998	995	987	972	943	896	829	739	631	513	395	287	196	126	076	042	022	011	005	002	001

[a] Initial decimal point has been omitted.

Source: From E.B. Roessler et al. 1978. J. Food Sci. 43(3):942–943.

Statistical Chart 3 Two-sample test, probability of x or more agreeing judgments in n trials (two-tailed, $p = 1/2$)[a]

n \ x	3	4	5	6	7	8	9	10	11	12	13	14	15	16	17	18	19	20	21	22	23	24	25	26	27	28	29	30	31	32	33	34	35	36	37
5	625	312	062																																
6		688	219	031																															
7			453	125	016																														
8			727	289	070	008																													
9				508	180	039	004																												
10				754	344	109	021	002																											
11					549	227	065	011	001																										
12					774	388	146	039	006																										
13						581	267	092	022	003																									
14						791	424	180	057	013	002																								
15							607	302	118	035	007	001																							
16							804	454	210	077	021	004	001																						
17								629	332	143	049	013	002																						
18								815	481	238	096	031	008	001																					
19									648	359	167	064	019	004	001																				
20									824	503	263	115	041	012	003																				
21										664	383	189	078	027	007	001																			
22										832	523	286	134	052	017	004	001																		
23											678	405	210	093	035	011	003																		
24											839	541	302	152	064	023	007	002																	
25												690	424	230	108	043	015	004	001																
26												845	557	327	169	076	029	009	002	001															
27													701	442	248	122	052	019	006	002															
28													851	572	345	185	087	036	013	004	001														
29														711	458	265	136	061	024	008	002	001													
30														856	585	362	200	099	043	016	005	001													
31															720	473	281	150	071	030	011	003	001												
32															860	597	377	215	100	050	020	007	002	001											
33																728	487	296	163	080	035	014	005	001											
34																864	608	392	229	121	058	024	009	003	001										
35																	736	500	310	175	090	041	017	006	002										
36																	868	681	405	243	132	065	029	011	004	001									
37																		743	511	324	188	099	047	020	008	003	001								
38																		871	627	418	256	143	073	034	014	005	002								
39																			749	522	337	200	108	053	024	009	003	001							
40																			875	636	430	268	154	081	038	017	006	002	001						
41																				755	533	349	211	117	060	028	012	004	001						
42																				878	644	441	280	164	088	044	020	008	003	001					
43																					761	542	360	222	126	066	032	014	005	002	001				
44																					880	652	451	291	174	096	049	023	010	004	001				
45																						766	551	371	233	135	072	036	016	007	002	001			
46																						883	659	461	302	184	104	054	026	011	005	002	001		
47																							771	560	382	243	144	079	040	019	008	003	001		
48																							885	665	471	312	193	111	069	029	013	006	002	001	
49																								775	568	392	253	152	085	044	021	009	004	001	
50																								888	672	480	322	203	119	065	033	015	007	003	001

[a] Initial decimal point has been omitted.

Source: From E.B. Roessler et al. 1978. J. Food Sci. 43(3):942–943. © Institute of Food Technologists. Reprinted with permission of the publisher.

Statistical Chart 4 Two-sample test, probability of *x* or more correct judgments in *n* trials (one-tailed, $p = 1/2$)[a]

n \ x	0	1	2	3	4	5	6	7	8	9	10	11	12	13	14	15	16	17	18	19	20	21	22	23	24	25	26	27	28	29	30	31	32	33	34	35	36
5		969	812	500	188	031																															
6		984	891	656	344	109	016																														
7		992	938	773	500	227	062	008																													
8		996	965	855	637	363	145	035	004																												
9		998	980	910	746	500	254	090	020	002																											
10		999	989	945	828	623	377	172	055	011	001																										
11			994	967	887	726	500	274	113	033	006																										
12			997	981	927	806	613	387	194	073	019	003																									
13			998	989	954	867	709	500	291	133	046	011	002																								
14			999	994	971	910	788	605	395	212	090	029	006	001																							
15				996	982	941	849	696	500	304	151	059	018	004																							
16				998	989	962	895	773	598	402	227	105	038	011	002																						
17				999	994	975	928	834	685	500	315	166	072	025	006	001																					
18				999	996	985	952	881	760	593	407	240	119	048	015	004	001																				
19					998	990	968	916	820	676	500	324	180	084	032	010	002																				
20					999	994	979	942	868	748	588	412	252	132	058	021	006	001																			
21					999	996	987	961	905	808	668	500	332	192	095	039	013	004	001																		
22						998	992	974	933	857	738	584	416	262	143	067	026	008	002																		
23						999	995	983	953	895	798	661	500	339	202	105	047	017	005	001																	
24						999	997	989	968	924	846	729	581	419	271	154	076	032	011	003	001																
25							998	993	978	946	885	788	655	500	345	212	115	054	022	007	002																
26							999	995	986	962	916	837	721	577	423	279	163	084	038	014	005	001															
27							999	997	990	974	939	876	779	649	500	351	221	124	061	026	010	003	001														
28								998	994	982	956	908	828	714	575	425	286	172	092	044	018	006	002														
29								999	996	988	969	932	868	771	644	500	356	229	132	068	031	012	004	001													
30								999	997	992	979	951	900	819	708	572	428	292	181	100	049	021	008	003	001												
31									998	995	985	965	925	859	763	640	500	360	237	141	075	035	015	005	002												
32									999	997	990	975	945	892	811	702	570	430	298	189	108	055	025	010	004	001											
33									999	998	993	982	960	919	852	757	636	500	364	243	148	081	040	018	007	002											
34										999	995	988	971	939	885	804	696	568	432	304	196	115	061	029	012	005	002										
35										999	997	992	980	955	912	845	750	632	500	368	250	155	088	045	020	008	003	001									
36										999	998	994	986	967	934	879	797	691	566	434	309	203	121	066	033	014	006	002	001								
37											999	996	990	976	951	906	838	744	629	500	371	256	162	094	049	024	010	004	001								
38											999	997	993	983	964	928	872	791	686	564	436	314	209	128	072	036	017	007	003	001							
39											999	998	995	988	973	946	900	832	739	625	500	375	261	168	100	054	027	012	005	002	001						
40												999	997	992	981	960	923	866	785	682	563	437	318	215	134	077	040	019	008	003	001						
41												999	998	994	986	970	941	894	826	734	622	500	378	266	174	106	059	030	014	006	002	001					
42													999	996	990	978	956	918	860	780	678	561	439	322	220	140	082	044	022	010	004	001					
43													999	997	993	984	967	937	889	820	729	620	500	380	271	180	111	063	033	016	007	003	001				
44													999	998	995	989	976	952	913	854	774	674	560	440	326	226	146	087	048	024	011	005	002	001			
45														999	997	992	982	964	932	884	814	724	617	500	383	276	186	116	068	036	018	008	003	001			
46														999	998	994	987	973	948	908	849	769	671	558	442	329	231	151	092	052	027	013	006	002	001		
47														999	998	996	991	980	961	928	879	809	720	615	500	385	280	191	121	072	039	020	009	004	002	001	
48															999	997	993	985	970	944	903	844	765	667	557	443	333	235	156	097	056	030	015	007	003	001	
49															999	998	995	989	978	957	924	874	804	716	612	500	388	284	196	126	076	043	022	012	005	002	001
50																999	997	992	984	968	941	899	839	760	664	556	444	336	240	161	101	059	032	016	008	003	001

[a] Initial decimal point has been omitted.

Source: From E.B. Roessler et al. 1978. J. Food Sci. 43(3):942–943. © Institute of Food Technologists. Reprinted with permission of the publisher.

Statistical Chart 5 Probability of x or more correct judgments in n trials of a two-out-of-five test (one-tailed, P = 0.1)[a]

n \ x	1	2	3	4	5	6	7	8	9	10	11	12	13
2	190	010											
3	271	028	001										
4	344	052	004										
5	410	081	009										
6	469	114	016	001									
7	522	150	026	003									
8	570	187	038	005									
9	613	225	053	008	001								
10	651	264	070	013	002								
11	686	303	090	019	003								
12	718	341	111	026	004	001							
13	746	379	134	034	006	001							
14	771	415	158	044	009	001							
15	794	451	184	056	013	002							
16	815	485	211	068	017	003	001						
17	833	518	238	083	022	005	001						
18	850	550	266	098	028	006	001						
19	865	580	295	115	035	009	002						
20	878	608	323	133	043	011	002						
21	891	635	352	152	052	014	003	001					
22	902	661	380	172	062	018	004	001					
23	911	685	408	193	073	023	006	001					
24	920	708	436	214	085	028	007	002					
25	928	729	463	236	098	033	009	002					
26	935	749	489	259	112	040	012	003	001				
27	942	767	515	282	127	047	015	004	001				
28	948	785	541	305	142	055	018	005	001				
29	953	801	565	329	158	064	022	006	002				
30	958	816	589	353	175	073	026	008	002				
31	962	830	611	376	193	083	031	010	003	001			
32	966	844	633	400	211	094	036	012	003	001			
33	969	856	654	423	230	106	042	014	004	001			
34	972	867	674	446	250	119	048	017	005	001			
35	975	878	694	469	269	132	055	020	006	002			
36	977	887	712	491	289	145	063	024	008	002	001		
37	980	896	730	514	309	160	071	027	009	003	001		
38	982	905	746	535	330	175	080	032	011	003	001		
39	984	912	762	556	350	190	089	037	013	004	001		
40	985	920	777	577	371	206	100	042	015	005	001		
41	987	926	791	597	392	223	110	048	018	006	002		
42	988	932	805	616	412	240	121	054	021	007	002	001	
43	989	938	818	635	433	257	133	061	024	009	003	001	
44	990	943	830	653	453	274	146	068	028	010	003	001	
45	991	948	841	671	473	292	159	076	032	012	004	001	
46	992	952	852	688	493	310	172	084	036	014	005	002	
47	993	956	862	704	512	329	186	093	041	016	006	002	001
48	994	960	871	720	531	347	200	102	046	019	007	002	001
49	994	963	880	735	550	365	215	112	052	022	008	003	001
50	995	966	888	750	569	384	230	122	058	025	009	003	001

[a] Initial decimal point has been omitted.

Statistical Chart 6 Upper α probability points of χ^2 distribution (entries are $\chi^2_{\alpha:\nu}$)

Instructions: (1) Enter the row of the table corresponding to the number of degrees of freedom (ν) for χ^2.
(2) Pick the value of χ^2 in that row from the column that corresponds to the predetermined proportional α – level.

$\nu \backslash \alpha$	**0.995**	**0.990**	**0.975**	**0.950**	**0.900**	**0.750**	**0.500**	**0.250**	**0.100**	**0.050**	**0.025**	**0.010**	**0.005**
1	0.0000393	0.000157	0.000982	0.00393	0.0158	0.102	0.455	1.32	2.71	3.84	5.02	6.63	7.88
2	0.0100	0.0201	0.0506	0.103	0.211	0.575	1.39	2.77	4.61	5.99	7.38	9.21	10.6
3	0.0717	0.115	0.216	0.352	0.584	1.21	2.37	4.11	6.25	7.81	9.35	11.3	12.8
4	0.207	0.297	0.484	0.711	1.06	1.92	3.36	5.39	7.78	9.49	11.1	13.3	14.9
5	0.412	0.554	0.831	1.15	1.61	2.67	4.35	6.63	9.24	11.1	12.8	15.1	16.7
6	0.676	0.872	1.24	1.64	2.20	3.45	5.35	7.84	10.6	12.6	14.4	16.8	18.5
7	0.989	1.24	1.69	2.17	2.83	4.25	6.35	9.04	12.0	14.1	16.0	18.5	20.3
8	1.34	1.65	2.18	2.73	3.49	5.07	7.34	10.2	13.4	15.5	17.5	20.1	22.0
9	1.73	2.09	2.70	3.33	4.17	5.90	8.34	11.4	14.7	16.9	19.0	21.7	23.6
10	2.16	2.56	3.25	3.94	4.87	6.74	9.34	12.5	16.0	18.3	20.5	23.2	25.2
11	2.60	3.05	3.82	4.57	5.58	7.58	10.3	13.7	17.3	19.7	21.9	24.7	26.8
12	3.07	3.57	4.40	5.23	6.30	8.44	11.3	14.8	18.5	21.0	23.3	26.2	28.3
13	3.57	4.11	5.01	5.89	7.04	9.30	12.3	16.0	19.8	22.4	24.7	27.7	29.8
14	4.07	4.66	5.63	6.57	7.79	10.2	13.3	17.1	21.1	23.7	26.1	29.1	31.3
15	4.60	5.23	6.26	7.26	8.55	11.0	14.3	18.2	22.3	25.0	27.5	30.6	32.8
16	5.14	5.81	6.91	7.96	9.31	11.9	15.3	19.4	23.5	26.3	28.8	32.0	34.3
17	5.70	6.41	7.56	8.67	10.1	12.8	16.3	20.5	24.8	27.6	30.2	33.4	35.7
18	6.26	7.01	8.23	9.39	10.9	13.7	17.3	21.6	26.0	28.9	31.5	34.8	37.2
19	6.84	7.63	8.91	10.1	11.7	14.6	18.3	22.7	27.2	30.1	32.9	36.2	38.6
20	7.43	8.26	9.59	10.9	12.4	15.5	19.3	23.8	28.4	31.4	34.2	37.6	40.0
21	8.03	8.90	10.3	11.6	13.2	16.3	20.3	24.9	29.6	32.7	35.5	38.9	41.4
22	8.64	9.54	11.0	12.3	14.0	17.2	21.3	26.0	30.8	33.9	36.8	40.3	42.8
23	9.26	10.2	11.7	13.1	14.8	18.1	22.3	27.1	32.0	35.2	38.1	41.6	44.2
24	9.89	10.9	12.4	13.8	15.7	19.0	23.3	28.2	33.2	36.4	39.4	43.0	45.6
25	10.5	11.5	13.1	14.6	16.5	19.9	24.3	29.3	34.4	37.7	40.6	44.3	46.9
26	11.2	12.2	13.8	15.4	17.3	20.8	25.3	30.4	35.6	38.9	41.9	45.6	48.3
27	11.8	12.9	14.6	16.2	18.1	21.7	26.3	31.5	36.7	40.1	43.2	47.0	49.6
28	12.5	13.6	15.3	16.9	18.9	22.7	27.3	32.6	37.9	41.3	44.5	48.3	51.0
29	13.1	14.3	16.0	17.7	19.8	23.6	28.3	33.7	39.1	42.6	45.7	49.6	52.3
30	13.8	15.0	16.8	18.5	20.6	24.5	29.3	34.8	40.3	43.8	47.0	50.9	53.7

Source: Reprinted with permission from Meilgaard, M.; Civille, G. Vance; Carr, Thomas B. 1987. Sensory evaluation techniques. Vol. II. CRC Press, Inc., Boca Raton, Fla. 159 pp.

Statistical Chart 7 Significant studentized range at the 5% level

Degrees of freedom	Number of treatments, *a*																		
f	2	3	4	5	6	7	8	9	10	11	12	13	14	15	16	17	18	19	20
1	18.0	26.7	32.8	37.2	40.5	43.1	45.4	47.3	49.1	50.6	51.9	53.2	54.3	55.4	56.3	57.2	58.0	58.8	59.6
2	6.09	8.28	9.80	10.89	11.73	12.43	13.03	13.54	13.99	14.39	14.75	15.08	15.38	15.65	15.91	16.14	16.36	16.57	16.77
3	4.50	5.88	6.83	7.51	8.04	8.47	8.85	9.18	9.46	9.72	9.95	10.16	10.35	10.52	10.69	10.84	10.98	11.12	11.24
4	3.93	5.00	5.76	6.31	6.73	7.06	7.35	7.60	7.83	8.03	8.21	8.37	8.52	8.67	8.80	8.92	9.03	9.14	9.24
5	3.61	4.54	5.18	5.64	5.99	6.28	6.52	6.74	6.93	7.10	7.25	7.39	7.52	7.64	7.75	7.86	7.95	8.04	8.13
6	3.46	4.34	4.90	5.31	5.63	5.89	6.12	6.32	6.49	6.65	6.79	6.92	7.04	7.14	7.24	7.34	7.43	7.51	7.59
7	3.34	4.16	4.68	5.06	5.35	5.59	5.80	5.99	6.15	6.29	6.42	6.54	6.65	6.75	6.84	6.93	7.01	7.08	7.16
8	3.26	4.04	4.53	4.89	5.17	5.40	5.60	5.77	5.92	6.05	6.18	6.29	6.39	6.48	6.57	6.65	6.73	6.80	6.87
9	3.20	3.95	4.42	4.76	5.02	5.24	5.43	5.60	5.74	5.87	5.98	6.09	6.19	6.28	6.36	6.44	6.51	6.58	6.65
10	3.15	3.88	4.33	4.66	4.91	5.12	5.30	5.46	5.60	5.72	5.83	5.93	6.03	6.12	6.20	6.27	6.34	6.41	6.47
11	3.11	3.82	4.26	4.58	4.82	5.03	5.20	5.35	5.49	5.61	5.71	5.81	5.90	5.98	6.06	6.14	6.20	6.27	6.33
12	3.08	3.77	4.20	4.51	4.75	4.95	5.12	5.27	5.40	5.51	5.61	5.71	5.80	5.88	5.95	6.02	6.09	6.15	6.21
13	3.06	3.73	4.15	4.46	4.69	4.88	5.05	5.19	5.32	5.43	5.53	5.63	5.71	5.79	5.86	5.93	6.00	6.06	6.11
14	3.03	3.70	4.11	4.41	4.64	4.83	4.99	5.13	5.25	5.36	5.46	5.56	5.64	5.72	5.79	5.86	5.92	5.98	6.03
15	3.01	3.67	4.08	4.37	4.59	4.78	4.94	5.08	5.20	5.31	5.40	5.49	5.57	5.65	5.72	5.79	5.85	5.91	5.96
16	3.00	3.65	4.05	4.34	4.56	4.74	4.90	5.03	5.15	5.26	5.35	5.44	5.52	5.59	5.66	5.73	5.79	5.84	5.90
17	2.98	3.62	4.02	4.31	4.52	4.70	4.86	4.99	5.11	5.21	5.31	5.39	5.47	5.55	5.61	5.68	5.74	5.79	5.84
18	2.97	3.61	4.00	4.28	4.49	4.67	4.83	4.96	5.07	5.17	5.27	5.35	5.43	5.50	5.57	5.63	5.69	5.74	5.79
19	2.96	3.59	3.98	4.26	4.47	4.64	4.79	4.92	5.04	5.14	5.23	5.32	5.39	5.46	5.53	5.59	5.65	5.70	5.75
20	2.95	3.58	3.96	4.24	4.45	4.62	4.77	4.90	5.01	5.11	5.20	5.28	5.36	5.43	5.50	5.56	5.61	5.66	5.71
24	2.92	3.53	3.90	4.17	4.37	4.54	4.68	4.81	4.92	5.01	5.10	5.18	5.25	5.32	5.38	5.44	5.50	5.55	5.99
30	2.89	3.48	3.84	4.11	4.30	4.46	4.60	4.72	4.83	4.92	5.00	5.08	5.15	5.21	5.27	5.33	5.38	5.43	5.48
40	2.86	3.44	3.79	4.04	4.23	4.39	4.52	4.63	4.74	4.82	4.90	4.98	5.05	5.11	5.17	5.22	5.27	5.32	5.36
60	2.83	3.40	3.74	3.98	4.16	4.31	4.44	4.55	4.65	4.73	4.81	4.88	4.94	5.00	5.06	5.11	5.15	5.20	5.24
120	2.80	3.36	3.69	3.92	4.10	4.24	4.36	4.47	4.56	4.64	4.71	4.78	4.84	4.90	4.95	5.00	5.04	5.09	5.13
∞	2.77	3.32	3.63	3.86	4.03	4.17	4.29	4.39	4.47	4.55	4.62	4.68	4.74	4.80	4.84	4.89	4.93	4.97	5.01

Source: Reprinted by permission from Snedecor, G.W.; Cochran, W.G. 1989. Statistical methods. 8th edition. © Iowa State University Press, Ames, Iowa.

Statistical Chart 8 Distribution of *t*

Degrees of freedom	Probability of a larger value, sign ignored								
f	0.500	0.400	0.200	0.100	0.050	0.025	0.010	0.005	0.001
1	1.000	1.376	3.078	6.314	12.706	25.452	63.657		
2	0.816	1.061	1.886	2.920	4.303	6.205	9.925	14.089	31.598
3	0.765	0.978	1.638	2.353	3.182	4.176	5.841	7.453	12.941
4	0.741	0.941	1.533	2.132	2.776	3.495	4.604	5.598	8.610
5	0.727	0.920	1.476	2.015	2.571	3.163	4.032	4.773	6.859
6	0.718	0.906	1.440	1.943	2.447	2.969	3.707	4.317	5.959
7	0.711	0.896	1.415	1.895	2.365	2.841	3.499	4.029	5.405
8	0.706	0.889	1.397	1.860	2.306	2.752	3.355	3.832	5.041
9	0.703	0.883	1.383	1.833	2.262	2.685	3.250	3.690	4.781
10	0.700	0.879	1.372	1.812	2.228	2.634	3.169	3.581	4.587
11	0.697	0.876	1.363	1.796	2.201	2.593	3.106	3.497	4.437
12	0.695	0.873	1.356	1.782	2.179	2.560	3.055	3.428	4.318
13	0.694	0.870	1.350	1.771	2.160	2.533	3.012	3.372	4.221
14	0.692	0.868	1.345	1.761	2.145	2.510	2.977	3.326	4.140
15	0.691	0.866	1.341	1.753	2.131	2.490	2.947	3.286	4.073
16	0.690	0.865	1.337	1.746	2.120	2.473	2.921	3.252	4.015
17	0.689	0.863	1.333	1.740	2.110	2.458	2.898	3.222	3.965
18	0.688	0.862	1.330	1.734	2.101	2.445	2.878	3.197	3.922
19	0.688	0.861	1.328	1.729	2.093	2.433	2.861	3.174	3.883
20	0.687	0.860	1.325	1.725	2.086	2.423	2.845	3.153	3.850
21	0.686	0.859	1.323	1.721	2.080	2.414	2.831	3.135	3.819
22	0.686	0.858	1.321	1.717	2.074	2.406	2.819	3.119	3.792
23	0.685	0.858	1.319	1.714	2.069	2.398	2.807	3.104	3.767
24	0.685	0.857	1.318	1.711	2.064	2.391	2.797	3.090	3.745
25	0.684	0.856	1.316	1.708	2.060	2.385	2.787	3.078	3.725
26	0.684	0.856	1.315	1.706	2.056	2.379	2.779	3.067	3.707
27	0.684	0.855	1.314	1.703	2.052	2.373	2.771	3.056	3.690
28	0.683	0.855	1.313	1.701	2.048	2.368	2.763	3.047	3.674
29	0.683	0.854	1.311	1.699	2.045	2.364	2.756	3.038	3.659
30	0.683	0.854	1.310	1.697	2.042	2.360	2.750	3.030	3.646
35	0.682	0.852	1.306	1.690	2.030	2.342	2.724	2.996	3.591
40	0.681	0.851	1.303	1.684	2.021	2.329	2.704	2.971	3.551
45	0.680	0.850	1.301	1.680	2.014	2.319	2.690	2.952	3.520
50	0.680	0.849	1.299	1.676	2.008	2.310	2.678	2.937	3.496
55	0.679	0.849	1.297	1.673	2.004	2.304	2.669	2.925	3.476
60	0.679	0.848	1.296	1.671	2.000	2.299	2.660	2.915	3.460
70	0.678	0.847	1.294	1.667	1.994	2.290	2.648	2.899	3.435
80	0.678	0.847	1.293	1.665	1.989	2.284	2.638	2.887	3.416
90	0.678	0.846	1.291	1.662	1.986	2.279	2.631	2.878	3.402
100	0.677	0.846	1.290	1.661	1.982	2.276	2.625	2.871	3.390
120	0.677	0.845	1.289	1.658	1.980	2.270	2.617	2.860	3.373
∞	0.6745	0.8416	1.2816	1.6448	1.9600	2.2414	2.5758	2.8070	3.2905

Source: Reprinted by permission from Snedecor, G.W.; Cochran, W.G. 1989. Statistical methods, 8th edition. © Iowa State University Press, Ames, Iowa.

Statistical Chart 9 Variance ratio—5 percent points for distribution of F
n_1—degrees of freedom for numerator
n_2—degrees of freedom for denominator

n_2 \ n_1	1	2	3	4	5	6	8	12	24	∞
1	161.4	199.5	215.7	224.6	230.2	234.0	238.9	243.9	249.0	254.3
2	18.51	19.00	19.16	19.25	19.30	19.33	19.37	19.41	19.45	19.50
3	10.13	9.55	9.28	9.12	9.01	8.94	8.84	8.74	8.64	8.53
4	7.71	6.94	6.59	6.39	6.26	6.16	6.04	5.91	5.77	5.63
5	6.61	5.79	5.41	5.19	5.05	4.95	4.82	4.68	4.53	4.36
6	5.99	5.14	4.76	4.53	4.39	4.28	4.15	4.00	3.84	3.67
7	5.59	4.74	4.35	4.12	3.97	3.87	3.73	3.57	3.41	3.23
8	5.32	4.46	4.07	3.84	3.69	3.58	3.44	3.28	3.12	2.93
9	5.12	4.26	3.86	3.63	3.48	3.37	3.23	3.07	2.90	2.71
10	4.96	4.10	3.71	3.48	3.33	3.22	3.07	2.91	2.74	2.54
11	4.84	3.98	3.59	3.36	3.20	3.09	2.95	2.79	2.61	2.40
12	4.75	3.88	3.49	3.26	3.11	3.00	2.85	2.69	2.50	2.30
13	4.67	3.80	3.41	3.18	3.02	2.92	2.77	2.60	2.42	2.21
14	4.60	3.74	3.34	3.11	2.96	2.85	2.70	2.53	2.35	2.13
15	4.54	3.68	3.29	3.06	2.90	2.79	2.64	2.48	2.29	2.07
16	4.49	3.63	3.24	3.01	2.85	2.74	2.59	2.42	2.24	2.01
17	4.45	3.59	3.20	2.96	2.81	2.70	2.55	2.38	2.19	1.96
18	4.41	3.55	3.16	2.93	2.77	2.66	2.51	2.34	2.15	1.92
19	4.38	3.52	3.13	2.90	2.74	2.63	2.48	2.31	2.11	1.88
20	4.35	3.49	3.10	2.87	2.71	2.60	2.45	2.28	2.08	1.84
21	4.32	3.47	3.07	2.84	2.68	2.57	2.42	2.25	2.05	1.81
22	4.30	3.44	3.05	2.82	2.66	2.55	2.40	2.23	2.03	1.78
23	4.28	3.42	3.03	2.80	2.64	2.53	2.38	2.20	2.00	1.76
24	4.26	3.40	3.01	2.78	2.62	2.51	2.36	2.18	1.98	1.73
25	4.24	3.38	2.99	2.76	2.60	2.49	2.34	2.16	1.96	1.71
26	4.22	3.37	2.98	2.74	2.59	2.47	2.32	2.15	1.95	1.69
27	4.21	3.35	2.96	2.73	2.57	2.46	2.30	2.13	1.93	1.67
28	4.20	3.34	2.95	2.71	2.56	2.44	2.29	2.12	1.91	1.65
29	4.18	3.33	2.93	2.70	2.54	2.43	2.28	2.10	1.90	1.64
30	4.17	3.32	2.92	2.69	2.53	2.42	2.27	2.09	1.89	1.62
40	4.08	3.23	2.84	2.61	2.45	2.34	2.18	2.00	1.79	1.51
60	4.00	3.15	2.76	2.52	2.37	2.25	2.10	1.92	1.70	1.39
120	3.92	3.07	2.68	2.45	2.29	2.17	2.02	1.83	1.61	1.25
∞	3.84	2.99	2.60	2.37	2.21	2.09	1.94	1.75	1.52	1.00

(continued)

Statistical Chart 9 (*concluded*) Variance ratio—1 percent points for distribution of F
n_1—degrees of freedom for numerator
n_2—degrees of freedom for denominator

n_2 \ n_1	1	2	3	4	5	6	8	12	24	∞
1	4052	4999	5403	5625	5764	5859	5981	6106	6234	6366
2	98.49	99.00	99.17	99.25	99.30	99.33	99.36	99.42	99.46	99.50
3	34.12	30.81	29.46	28.71	28.24	27.91	27.49	27.05	26.60	26.12
4	21.20	18.00	16.69	15.98	15.52	15.21	14.80	14.37	13.93	13.46
5	16.46	13.27	12.06	11.39	10.97	10.67	10.29	9.89	9.47	9.02
6	13.74	10.92	9.78	9.15	8.75	8.47	8.10	7.72	7.31	6.88
7	12.25	9.55	8.45	7.85	7.46	7.19	6.84	6.47	6.07	5.65
8	11.26	8.65	7.59	7.01	6.63	6.37	6.03	5.67	5.28	4.86
9	10.56	8.02	6.99	6.42	6.06	5.80	5.47	5.11	4.73	4.31
10	10.04	7.56	6.55	5.99	5.64	5.39	5.06	4.71	4.33	3.91
11	9.65	7.20	6.22	5.67	5.32	5.07	4.74	4.40	4.02	3.60
12	9.33	6.93	5.95	5.41	5.06	4.82	4.50	4.16	3.78	3.36
13	9.07	6.70	5.74	5.20	4.86	4.62	4.30	3.96	3.59	3.16
14	8.86	6.51	5.56	5.03	4.69	4.46	4.14	3.80	3.43	3.00
15	8.68	6.36	5.42	4.89	4.56	4.32	4.00	3.67	3.29	2.87
16	8.53	6.23	5.29	4.77	4.44	4.20	3.89	3.55	3.18	2.75
17	8.40	6.11	5.18	4.67	4.34	4.10	3.79	3.45	3.08	2.65
18	8.28	6.01	5.09	4.58	4.25	4.01	3.71	3.37	3.00	2.57
19	8.18	5.93	5.01	4.50	4.17	3.94	3.63	3.30	2.92	2.49
20	8.10	5.85	4.94	4.43	4.10	3.87	3.56	3.23	2.86	2.42
21	8.02	5.78	4.87	4.37	4.04	3.81	3.51	3.17	2.80	2.36
22	7.94	5.72	4.82	4.31	3.99	3.76	3.45	3.12	2.75	2.31
23	7.88	5.66	4.76	4.26	3.94	3.71	3.41	3.07	2.70	2.26
24	7.82	5.61	4.72	4.22	3.90	3.67	3.36	3.03	2.66	2.21
25	7.77	5.57	4.68	4.18	3.86	3.63	3.32	2.99	2.62	2.17
26	7.72	5.53	4.64	4.14	3.82	3.59	3.29	2.96	2.58	2.13
27	7.68	5.49	4.60	4.11	3.78	3.56	3.26	2.93	2.55	2.10
28	7.64	5.45	4.57	4.07	3.75	3.53	3.23	2.90	2.52	2.06
29	7.60	5.42	4.54	4.04	3.73	3.50	3.20	2.87	2.49	2.03
30	7.56	5.39	4.51	4.02	3.70	3.47	3.17	2.84	2.47	2.01
40	7.31	5.18	4.31	3.83	3.51	3.29	2.99	2.66	2.29	1.80
60	7.08	4.98	4.13	3.65	3.34	3.12	2.82	2.50	2.12	1.60
120	6.85	4.79	3.95	3.48	3.17	2.96	2.66	2.34	1.95	1.38
∞	6.64	4.60	3.78	3.32	3.02	2.80	2.51	2.18	1.79	1.00

Source: Table 9 is taken from Table V of Fisher and Yates: 1974 Statistical Tables for Biological, Agricultural and Medical Research published by Longman Group UK Ltd. London (previously published by Oliver and Boyd Ltd. Edinburgh) and by permission of the authors and publishers.

Statistical Chart 10 Table of *t* for one-sided Dunnett's test for comparing control against each of *p* other treatment means at the 5% level

	p, Number of treatment means (excluding the control)								
df	1	2	3	4	5	6	7	8	9
5	2.02	2.44	2.68	2.85	2.98	3.08	3.16	3.24	3.30
6	1.94	2.34	2.56	2.71	2.83	2.92	3.00	3.07	3.12
7	1.89	2.27	2.48	2.62	2.73	2.82	2.89	2.95	3.01
8	1.86	2.22	2.42	2.55	2.66	2.74	2.81	2.87	2.92
9	1.83	2.18	2.37	2.50	2.60	2.68	2.75	2.81	2.86
10	1.81	2.15	2.34	2.47	2.56	2.64	2.70	2.76	2.81
11	1.80	2.13	2.31	2.44	2.53	2.60	2.67	2.72	2.77
12	1.78	2.11	2.29	2.41	2.50	2.58	2.64	2.69	2.74
13	1.77	2.09	2.27	2.39	2.48	2.55	2.61	2.66	2.71
14	1.76	2.08	2.25	2.37	2.46	2.53	2.59	2.64	2.69
15	1.75	2.07	2.24	2.36	2.44	2.51	2.57	2.62	2.67
16	1.75	2.06	2.23	2.34	2.43	2.50	2.56	2.61	2.65
17	1.74	2.05	2.22	2.33	2.42	2.49	2.54	2.59	2.64
18	1.73	2.04	2.21	2.32	2.41	2.48	2.53	2.58	2.62
19	1.73	2.03	2.20	2.31	2.40	2.47	2.52	2.57	2.61
20	1.72	2.03	2.19	2.30	2.39	2.46	2.51	2.56	2.60
24	1.71	2.01	2.17	2.28	2.36	2.43	2.48	2.53	2.57
30	1.70	1.99	2.15	2.25	2.33	2.40	2.45	2.50	2.54
40	1.68	1.97	2.13	2.23	2.31	2.37	2.42	2.47	2.51
60	1.67	1.95	2.10	2.21	2.28	2.35	2.39	2.44	2.48
120	1.66	1.93	2.08	2.18	2.26	2.32	2.37	2.41	2.45
∞	1.64	1.92	2.06	2.16	2.23	2.29	2.34	2.38	2.42

Source: Reprinted with permission from Journal of American Statistical Association—Dunnett, Charles W. 1955. A multiple comparison procedure for comparing several treatments with a control. J. Am. Stat. Assoc. 50:1096–1121.

Statistical Chart 11 Table of *t* for two-sided Dunnett's test for comparing control against each of *p* other treatment means at the 5% level

df	1	2	3	4	5	6	7	8	9	10	11	12	15	20
	p, Number of treatment means (excluding the control)													
5	2.57	3.03	3.29	3.48	3.62	3.73	3.82	3.90	3.97	4.03	4.09	4.14	4.26	4.42
6	2.45	2.86	3.10	3.26	3.39	3.49	3.57	3.64	3.71	3.76	3.81	3.86	3.97	4.11
7	2.36	2.75	2.97	3.12	3.24	3.33	3.41	3.47	3.53	3.58	3.63	3.67	3.78	3.91
8	2.31	2.67	2.88	3.02	3.13	3.22	3.29	3.35	3.41	3.46	3.50	3.54	3.64	3.76
9	2.26	2.61	2.81	2.95	3.05	3.14	3.20	3.26	3.32	3.36	3.40	3.44	3.53	3.65
10	2.23	2.57	2.76	2.89	2.99	3.07	3.14	3.19	3.24	3.29	3.33	3.36	3.45	3.57
11	2.20	2.53	2.72	2.84	2.94	3.02	3.08	3.14	3.19	3.23	3.27	3.30	3.39	3.50
12	2.18	2.50	2.68	2.81	2.90	2.98	3.04	3.09	3.14	3.18	3.22	3.25	3.34	3.45
13	2.16	2.48	2.65	2.78	2.87	2.94	3.00	3.06	3.10	3.14	3.18	3.21	3.29	3.40
14	2.14	2.46	2.63	2.75	2.84	2.91	2.97	3.02	3.07	3.11	3.14	3.18	3.26	3.36
15	2.13	2.44	2.61	2.73	2.82	2.89	2.95	3.00	3.04	3.08	3.12	3.15	3.23	3.33
16	2.12	2.42	2.59	2.71	2.80	2.87	2.92	2.97	3.02	3.06	3.09	3.12	3.20	3.30
17	2.11	2.41	2.58	2.69	2.78	2.85	2.90	2.95	3.00	3.03	3.07	3.10	3.18	3.27
18	2.10	2.40	2.56	2.68	2.76	2.83	2.89	2.94	2.98	3.01	3.05	3.08	3.16	3.25
19	2.09	2.39	2.55	2.66	2.75	2.81	2.87	2.92	2.96	3.00	3.03	3.06	3.14	3.23
20	2.09	2.38	2.54	2.65	2.73	2.80	2.86	2.90	2.95	2.98	3.02	3.05	3.12	3.22
24	2.06	2.35	2.51	2.61	2.70	2.76	2.81	2.86	2.90	2.94	2.97	3.00	3.07	3.16
30	2.04	2.32	2.47	2.58	2.66	2.72	2.77	2.82	2.86	2.89	2.92	2.95	3.02	3.11
40	2.02	2.29	2.44	2.54	2.62	2.68	2.73	2.77	2.81	2.85	2.87	2.90	2.97	3.06
60	2.00	2.27	2.41	2.51	2.58	2.64	2.69	2.73	2.77	2.80	2.83	2.86	2.92	3.00
120	1.98	2.24	2.38	2.47	2.55	2.60	2.65	2.69	2.73	2.76	2.79	2.81	2.87	2.95
∞	1.96	2.21	2.35	2.44	2.51	2.57	2.61	2.65	2.69	2.72	2.74	2.77	2.83	2.91

Source: Reproduced from Dunnett, C.W. 1964. New tables for multiple comparisons with a control. Biometrics 20:482–491; with permission of The Biometric Society.

CONVERSION FACTORS

Multiply an imperial number by the conversion factor given to get its metric equivalent.

Divide a metric number by the conversion factor given to get its equivalent in imperial units.

Imperial units	Approximate conversion factor	Metric units	
Length			
inch	25	millimetre	(mm)
foot	30	centimetre	(cm)
yard	0.9	metre	(m)
mile	1.6	kilometre	(km)
Area			
square inch	6.5	square centimetre	(cm^2)
square foot	0.09	square metre	(m^2)
square yard	0.836	square metre	(m^2)
square mile	259	hectare	(ha)
acre	0.40	hectare	(ha)
Volume			
cubic inch	16	cubic centimetre	(cm^3, mL, cc)
cubic foot	28	cubic decimetre	(dm^3)
cubic yard	0.8	cubic metre	(m^3)
fluid ounce	28	millilitre	(mL)
pint	0.57	litre	(L)
quart	1.1	litre	(L)
gallon (Imp.)	4.5	litre	(L)
gallon (U.S.)	3.8	litre	(L)
Weight			
ounce	28	gram	(g)
pound	0.45	kilogram	(kg)
short ton (2000 lb)	0.9	tonne	(t)
Pressure			
pounds per square inch	6.9	kilopascal	(kPa)
Power			
horsepower	746	watt	(W)
	0.75	kilowatt	(kW)
Speed			
feet per second	0.30	metres per second	(m/s)
miles per hour	1.6	kilometres per hour	(km/h)
Agriculture			
gallons per acre	11.23	litres per hectare	(L/ha)
quarts per acre	2.8	litres per hectare	(L/ha)
pints per acre	1.4	litres per hectare	(L/ha)
fluid ounces per acre	70	millilitres per hectare	(mL/ha)
tons per acre	2.24	tonnes per hectare	(t/ha)
pounds per acre	1.12	kilograms per hectare	(kg/ha)
ounces per acre	70	grams per hectare	(g/ha)
plants per acre	2.47	plants per hectare	
Temperature			
degrees Fahrenheit	(°F − 32) x 0.56 = °C or °F = 1.8 (°C) + 32	degrees Celsius	(°C)